# BRICKLAYING

## Peter Cartwright

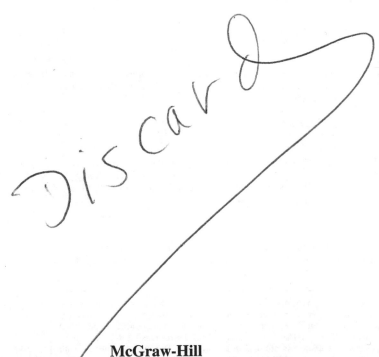
**McGraw-Hill**

New York • Chicago • San Francisco • Lisbon • London • Madrid
Mexico City • Milan • New Delhi • San Juan • Seoul
Singapore • Sydney • Toronto

**Cataloging-in-Publication Data is on file with the Library of Congress**

*McGraw-Hill*

*A Division of The McGraw·Hill Companies*

3 4 5 6 7 8 9 0   DOC/DOC   0 9 8 7 6 5 4 3

ISBN 0-07-139239-4

*The sponsoring editor for this book was Shelley Carr, the editing supervisor was Steven Melvin, and the production supervisor was Sherri Souffrance. It was set in Times by Wayne Palmer of McGraw-Hill Professional's Hightstown, N.J., composition unit.*

*Printed and bound by R. R. Donnelley & Sons Company.*

McGraw-Hill books are available at special quantity discounts to use as premiums and sales promotions, or for use in corporate training programs. For more information, please write to the Director of Special Sales, Professional Publishing, McGraw-Hill, Two Penn Plaza, New York, NY 10121-2298. Or contact your local bookstore.

 This book is printed on recycled, acid-free paper containing a minimum of 50% recycled, de-inked fiber.

# Contents

# Contents

*Contents*

# *Part II*

# Brickwork – Advanced Construction

**10. Introduction to Advanced Construction**

**11. Arches**

**12. Bay Windows**

**13. Setting up Profiles**

**14. Suspended Slab**

**15. Fireplace Construction**

# Contents

## Part IV

# Brickwork—Tips and Tricks

# Acknowledgments

• To Ronda, for being my wife, and the world's greatest Mum to our children and for the part I like most – *"being yourself."*

• To my Father, to a kind and loving gentleman, for his acts of love, wisdom, and encouragement, and his incredible teachings and intelligence, for which you will live forever.

• To my Father and Mother, who have given me knowledge and the ability to persist in many things.

*Special thanks to: My wife Ronda, Tom Cartwright, Noel & Sue Herley, Ross Coulter, Leonie Challacombe, Morris Maker, Technical Teacher Construction Rockhampton TAFE, and Jeff Barber. A sincere debt of gratitude is owed to the TAFE teachers, bricklayers, contractors, and laypeople who have provided me food for thought or assisted with information and questions.*

# Part I
# Bricklaying — The Basics
# 1
# Introduction

Bricklaying is an art. It is not merely laying one brick on another. Good bricklayers are highly skilled craftsmen sought after by building and bricklaying contractors.

There is as much precision in neat and tidy brickwork as in fine-tuning a car. Tradespeople should also be color conscious because of the huge variation in colors of bricks.

A good tradesperson never allows brick rubble and mortar to collect under his feet. Keep a clean working area, including mortar boards and scaffolding, and use tools correctly. Put them and use them in their place. Time spent looking for tools is time wasted, so always carry a tape and pencil handy, and know where to locate the correct tools when they are needed.

Always leave the mortar boards clean after a day's work—think of tomorrow! When working on a job place mortar boards in a position close and handy, and emphasize to the laborers how to work smart, not hard. Laborers and tradespeople who get to know each other's methods and abilities (or peculiarities) will both find their work easier.

This can't be overemphasized. I once watched a laborer stack bricks either side of his tradesperson, when it was obvious the tradesperson worked faster from his right and needed to slow down to use the bricks on the other pile, which the laborer kept replenishing!

Mortar too, should be put on boards where it will find the most use. Why carry bricks twice as far as you need to? Or mix more mortar because an ill-placed mortar-board was not used and allowed to dry?

Good tradespeople have pride in their work and because of their mastery of tools they are a joy to watch.

The efficiency of a bricklayer depends on the accuracy and neatness with which bricks can be laid. Speed can be gained through the use of fine-tuning and omitting all extra moves and the use of shortcuts and tricks of the trade that come with experience.

A good tradesperson can pick his or her bricks matching in color, and throw out bent bricks and bricks with cracks and flaws. Flawed bricks can ruin a job and mean more work in replacement for an unhappy customer. Decide early in your career that it is wiser to spend a few seconds laying a good brick than a half hour pulling it out and replacing it later.

Try not to pick up small amounts of mortar—the more the better (except corners and toothing where excess mortar could end up in the cavity). Increase your mortar-throwing ability to capacity to increase speed and efficiency when throwing out a bed. I aim to get four bricks buttered and bedded from one trowel of mortar, but don't heap mortar on the wall, because it's then harder to place the brick down to the line.

The excess mortar that comes out of the bed joint is wasted, and has to be double-handled when it falls on the ground or smothers the wall with compo stains. Worse still it could fill up the cavity or involve extra work and time in throwing it back on the compo board—all unnecessary, time-consuming movements.

I always maintain that a bricklayer's greatest attribute is his or her eye. With practice a person can get pretty good at "sighting" jobs. This is not to say levels are not important, but I have seen so many levels that for one reason or another were not accurate or differed from other levels. This has convinced me to trust my judgment and a good professional will work to develop this attribute.

Get into the habit of sighting along the courses of brickwork when finished, this leads to self-improvement.

Self-correction is the best way to learn to lay bricks.

Brickwork highlights the exterior of most houses and is not only a structural component but also protects weatherwise, and decorates.

It is probably one of the most inspected features of a house.

All brickwork should be checked frequently to ensure that it is plumb and level and laid to the line.

**Peter Cartwright 2002**

# Electricity

There are no second chances with electricity. Keep all cords out of wet areas. Be careful not to put scaffold legs on cords because when you load up the scaffold it will crush the cord. Plugging cords in the meter box and shutting the door is a sure way of cutting through the cords. Power cords into meter boxes can also be damaged as the lid vibrates and rattles as work, such as nailing in ties, takes place on adjoining walls. Either take the door off the meter box or tape it closed. Maybe tape some insulation over the cords where they enter the meter box.

When cords extend across the ground, run lengths of timber beside the cord in areas where there is constant traffic such as wheelbarrow or constant walking. Make sure there is a grounding unit on the site, particularly in renovation work, as old buildings may not be fitted with a ground.

# Bricks

There are generally two kinds of bricks

1. Pressed bricks — frog bricks (Fig. 1)

**Figure 1.1**

2. Extruded bricks — bricks with holes (Fig 1.2)

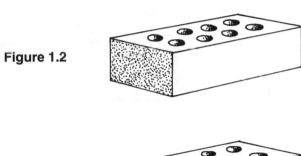

**Figure 1.2**

**Figure 1.3**
King closure

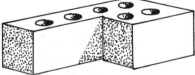

**Figure 1.4**
Queen closure

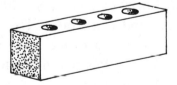

**Figure 1.5**
1/2 Bal

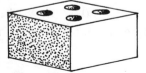

Some bricks require wetting due sometimes to the unavailability of good loam or sand. Bricks that are not silicon dipped require a good wetting before laying.

Symptoms that indicate that bricks require wetting are:

1. The drying out of mortar directly after the bed has been spread, and

2. Difficulty in tapping bricks down to the line.

Wetting the bricks retards the drying out of the mortar and allows easier raking out and cleaning.

I recommend all bent bricks be thrown out, but if this is impossible lay the bent bricks with the points up keeping them all laid the same way.

**Figure 1.6**

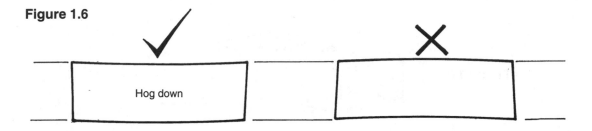

Hog down

# Mortar

Mortar is usually specified by the architect for engineered jobs or for heavy construction projects where variations in the usual ingredients of cement, sand, and water change the strength and setting characteristics of the mortar.

Personally I use a good strong mix of four shovels of loam to one of cement and a half shovel of lime mixed to consistency with water. Mixes should be consistent in color, composition, and workability. Use a container for measuring plasticizer or color that has to be added to the mix. *N.B. Too much plasticizer will destroy mortar strength.* Use only proven brands and the recommended quantities per batch. Too much water on top of a mix is an indicator that there is a lack of plasticizer. The use of a shovel of lime per batch enables you to cut back on plasticizer and adds tremendously to taking deadness out of colored mortar. It also makes the batch easier to mix and work with. If you must use additional plasticizer in the mortar, throw in an extra shovel of cement to bring the batch back to full strength.

Refer to Table 33.2 for mortar strength and ratio of sand to cement.

*Entanglement with rotating or moving machinery causes the worst accidents I have ever seen. Never put your hands or arms into a mixer — apart from the lime burns and the possibility of skin problems, the hazards are just too great.*

When we talk about *strength* what actually happens on a job site is another story. One big problem is ensuring mortar is mixed consistently and is fresh. When there are eight bricklayers screaming for mortar, sometimes the assistant could allow the mixes to be a little over or under, so here are my hints to ensure that your mortar mixes are always consistent and fresh.

First, the mixers should be at least a 3-cubic foot bowl. I have fitted specially designed beaters to my mixer bowl and I have increased the speed of the mixer up to 32 rpm (Fig. 1. 7).

Strengthening the bowl is also important, as is fitting double pulleys to any drive gear. This allows the batch to be properly mixed in a few minutes. I consider there is more to mixing mortar than meets the eye, and here is how I think it should be done.

**Figure 1.7**

*Step 1* First add the water then mix in the plasticizer with the water, which means you need less plasticizer because it will be mixed with the water and be carried through the other ingredients. On our jobs we use even less plasticizer because our mixer has beaters with the ability to mix the mortar really well, coating the sand with the cement and lime equally.

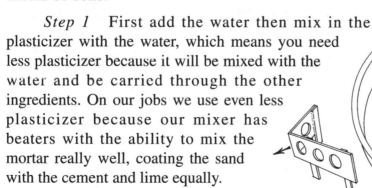

*Step 2*   If lime is to be used it should be added to the water/plasticizer mix. This has the effect of "killing" the lime. Adding water to lime in a dry mix can cause the lime to react and to spit back, with the potential to cause dangerous burning or serious eye injuries.

*Step 3*   Add 10 shovels of sand.

*Step 4*   Add half a 40-kg bag of cement. (Never put the cement into the water because it will stick to the inside of the bowl.)

*Step 5*   Add another 10 shovels of sand and leave it to mix for a minute.

As I said before we use a mixer no smaller than 3ft$^3$ to handle all these ingredients. As important, I always use a half a bag of cement as it's easy to remember and helps to ensure it always goes in the batch. I would have to say that I pity the poor assistant who makes a batch of mortar without cement and then proceeds to put it on the mortar boards. The abuse from the bricklayers would be far too much. Another good point about using a half bag of cement in each batch is that towards the end of the day when fatigue sets in and the shovels of sand are getting smaller the batch can only get stronger. I like this method because under extreme pressure we are eliminating the thinking.

Not to put too fine a point on things, I remember a time many years ago when part of the criteria for employing assistants was that they had to wear rubber thongs or they wouldn't get the job. This was because it was often said ten fingers, ten toes that bowl takes twenty shovels of sand.

# Tools

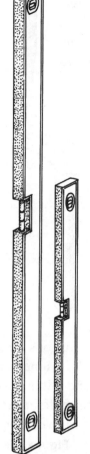

• **Long Spirit Level or 4'(1200 mm) Level**. The most common level/building corners and checking for level and plumb
**Figure 1.8**

• **Small Spirit Level or 2'(600 mm) Level**. Used in situations for speed and where the long level is awkward and clumsy, e.g., tight corners, piers, rakings on roofs

• **Brickie's Trowel**. Many trowels are manufactured, and are bought to the bricklayer's requirements and taste
**Figure 1.9**

• **Pointing Trowel**. Used in the parging in the gathering of a fireplace and the filling up of joints and cracks and also for struck pointing
**Figure 1.10**

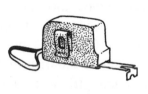

• **Tape**. Used in the checking of profiles at the beginning of a job, and frequently used in the constant checking in the requirements of bricklaying (windows and door heads)
**Figure 1.11**

• **Line and Pins**. Two pins with a line wound onto them. Used in the laying of wall bricks. Where the sills protrude out from the wall the pin is driven into the brickwork each end of the sill and the line set parallel with the two end sill bricks. Also used as a means to connect the line into an internal corner where it is impossible to use line blocks.
**Figure 1.12**

• **Stanley Knife**
**Figure 1.13**

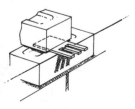

• **Tingle**. This is cut from a piece of metal. **Figure 1.14**

• **Plugging Chisel**. Used to chisel out the mortar bed and perp joints around the outside of a brick to be removed
**Figure 1.15**

• **Cold Chisel**. Used to remove any concrete or obstructing material in the path where bricks are to be laid
**Figure 1.16**

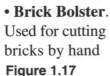

• **Brick Bolster**. Used for cutting bricks by hand
**Figure 1.17**

• **Lump Hammer**. Used to hit bolster or other chisels
**Figure 1.18**

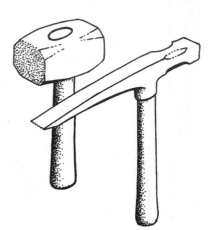

• **Rule**. Used for marking bricks to be cut with a bolster and marking bricks to be cut precisely to an arch
**Figure 1.19**

• **Scutch**. Used in the removal of protrusions on a brick after it has been cut — to make it a neater finish. Also in the fitting of pipes and wires into the brickwork where the occasional hole has to be scutched into the brick to fit over a tap
**Figure 1.20**

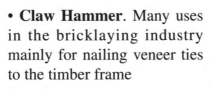

• **Claw Hammer**. Many uses in the bricklaying industry mainly for nailing veneer ties to the timber frame
**Figure 1.21**

• **Tin Snips.** **Figure 1.22**

• **Wheel Raker**. Tool for recessing the mortar joints around the brick face **Figure 1.23**

• **Rafter Square**. Helpful in setting out brick corners and checking brickwork for square
**Figure 1.24**

• **Gauge Rod**. Piece of aluminum 1"(25 mm) square with the appropriate gauge marked on it. Head height and all sill heights should be marked on the gauge rod
**Figure 1.25**

• **Line Blocks**. Generally made of timber and plastic and are the means by which the line is attached to a corner or profile
**Figure 1.26**

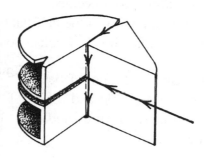

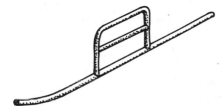

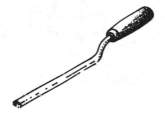

• **Round Iron Jointer**. Most commonly used in facing concrete blockwork. leaves a concave finish
**Figure 1.27**

• **Toothing Filler**. For the proper filling of toothings and also as a weep hole cleaner   **Figure 1.28**

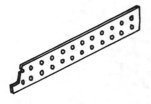

• **Corner or Tin Raker**. Cut from a piece of metal and used in the raking of corners —window heads, window and door jambs, and in tricky situations
**Figure 1.29**

# Bond

The most common bond is stretcher bond, referred to as half bond because each brick laps the bricks below by half a brick.

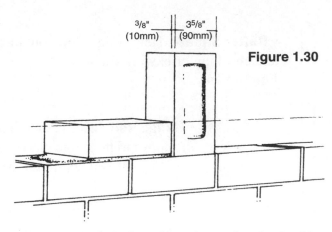

3/8"
(10mm)  3⁵/₈"
(90mm)

**Figure 1.30**

English bond is the strongest bond of all. It is possible to construct in single-skin English bond but this will mean an increase in labor costs because of the amount of cutting of the half-bat bricks. All varieties of bonds can be obtained by altering the two main bonds, English bond and Flemish bond.

All brickwork should be bonded where possible and all piers should have closures and ties to interlock them.

**Figure 1.31**   English bond

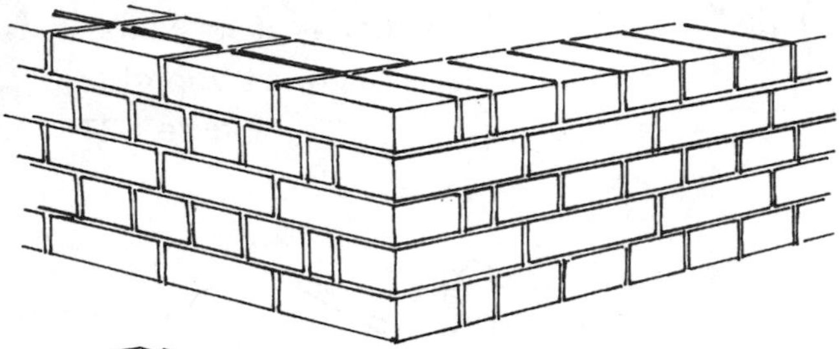

**Figure 1.32**   Stretcher bond

**Figure 1.33**

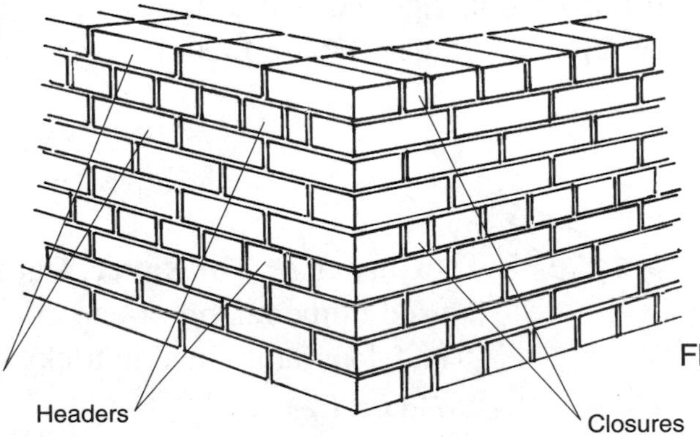

Stretchers

Headers

Closures

Flemish bond

**Figure 1.34**          Dutch bond

Bonding piers with queen closures prevents hair cracks between the piers and the wall, and extra strength is gained by cavity ties which prevent the pier from blowing out when filled with concrete.

Buttresses and nibs should be bonded for strength.

English Garden
Wall Bond

**Figure 1.35**

**Figure 1.36**          King closure

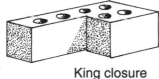

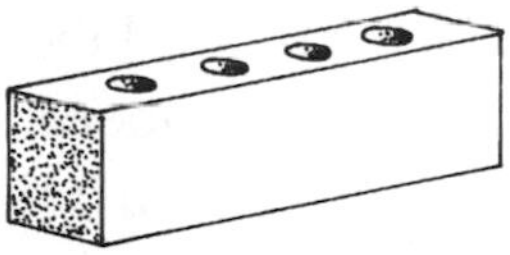

**Figure 1.37**          Queen closure

# Cutting Bricks with a Bolster or Saw

It is essential to keep the handle of the bolster ground clean, and the blade sharp, for a neater cut.

For right-handed cutters place the brick with the face upwards as shown.

Hold the bolster with the left hand as shown. Measure prescribed cut. Place a rule on the ground and hit the bolster with a sharp hard blow. Rotate the brick 180 degrees and repeat. Now turn the brick on its back and cut across the two side cuts. Scutch off excess protrusions.

**Figure 1.40**

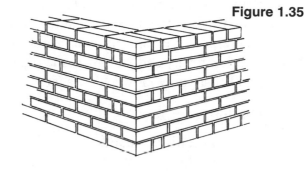

**Figure 1.38**

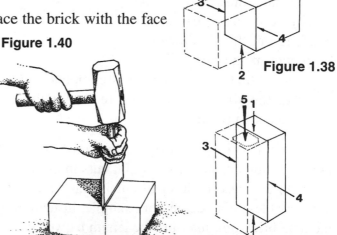

**Figure 1.39**

Difficult cuts can be made easier by placing a shovel of sand on the ground or scaffold as a cutting base to take the shock, or by first cutting 1½"–2"(35 – 50 mm) off the end of the brick. Now the final cut can be made. In most cases this will prevent the brick from breaking incorrectly, or shattering to pieces. Sawing bricks with power equipment is speedy and easy. A diamond-toothed saw is usually used at high speed. These saws rarely shatter or crack bricks, and leave a nice neat and clean edge.

If the cut is not too difficult, bricks can sometimes be cut through using a power saw fitted with a masonry cutting wheel. Eye protection should be used in all cases where power equipment is used for cutting. With diamond saws, take care to hold the brick firmly and the saw in the right position. Damage to the blades of these saws could cost you your week's wages.

### Guillotine

This machine, used correctly, can take a lot of pressure off the laborers, and with a clever operator could cut up to 100 bricks in 10 min. The trick is to write down the size of the cuts and the number of courses needed by each cut. Mark the size on a brick in pencil, place in the holder and adjust the size, and begin cutting. Even by working out the sills they can be precut at the beginning of

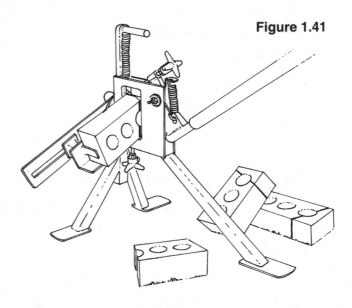

**Figure 1.41**

the job, so that when the wall is completed, the scaffold removed, and tide mark cleaned, the sills can be set up ready for the brickies, while the assistants rake, clean, and set up the scaffold.

When using cut bricks (broken bond), where possible place the cuts in inconspicuous places. I try to put cuts in an internal corner or under a window or a door. In the case of a small cut in the wall, cut small cuts off each brick so the change is less noticeable; for example, instead of a 4⅞" (125 mm) cut in a wall, take the next brick out, measure the length between:

13⅝" (345 mm) [4⅞" + ¾" + 8" (125 mm + 20 mm + 200 mm)]. The ¾" (20 mm) you add is the two perpendicular ends or the two lots of ⅜"(10 mm) of mortar; the 8" (200 mm) is the length of the brick plus one mortar joint or perpendicular; and the 4⅞" (125 mm) is our cut brick.

**Figure 1.42**

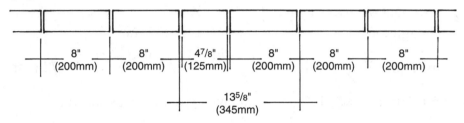

Take out 3 perpendicular joints — 30 mm: 13⅝" − 1⅛" = 12½" (345 − 30 = 315). Now divide by 2: 12½"/2 = 6¼" (315/2 = 158 mm). Two cuts of 6¼" (158 mm) is much better looking. The sketches explain this a little more clearly.

These are the formulae I use most.

**Figure 1.43**

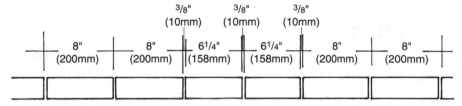

# Brick Formula 1

In cases of the unexpected half brick 4⅜" = 3⅝" + ¾" (110 = 90 + 20 mm) (2 perpends). Allow five bricks to be cut. Do this method in the least conspicuous place and take four bricks out, two each side of the 4⅜" bat.

**Figure 1.44**

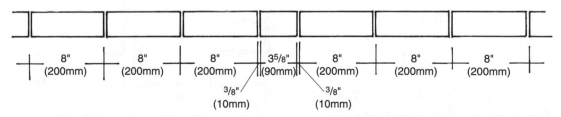

## Brick Formula 2

Measure distance between 4 × 8" = 32" + 4⅜" = 36⅜" (4 × 200 = 800 + 120 = 920 mm).

1. Total distance      X = 36⅜" (920 mm)
2. Total perpends to be deducted      Y = 2¼" (60 mm)
3. Amount of spaces required      Z = 5
4. Length of cut      L = 172 mm

For example. $\dfrac{X - Y}{Z} = L$    $\dfrac{36⅜ - 2¼}{5} = 6^{13}/_{16}\left(\dfrac{920 - 60}{5} = 172 \text{ mm}\right)$

Cut 5 bricks at 6¹³⁄₁₆" (172 mm) and lay appropriately as shown.

**Figure 1.45**

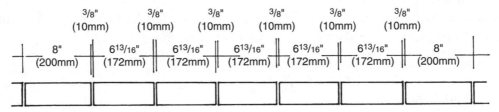

## Poor Bonding

This drawing shows an example where the brick bond is just 1³⁄₁₆" (30 mm) short of a window. In this case the lengthening of the perpends or the offsetting of the bond would destroy the look of the brickwork. I hope this diagram helps you one day to overcome this tricky situation.

For many years it has been thought that no bricks should be cut less than half a brick if laid in a wall. I don't think the great weight of experience supports this anymore. Personally I'd rather see bricks cut smaller up to an opening as opposed to opening up all the perpend joints.

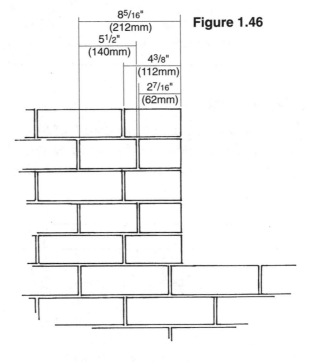

**Figure 1.46**

Make a judgment for yourself before you start, remembering that the early rules of the trade were applicable at a time when timber-framed windows were fitted. This has changed with the advent of aluminum-framed windows.

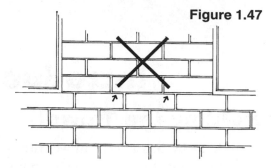

**Figure 1.47**

The figure above shows a brick wall between two windows where the perpends have been shifted out of vertical line to make the bricks work in between the windows.

**Figure 1.48a**

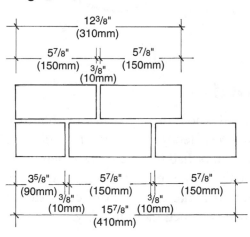

It's unsightly and unprofessional. It couldn't be called stretcher bond because stretcher bond is a half bond and in this example we have less than a quarter bond. At a minimum, it would be better to have at least one (1) reveal to work bond so cuts are only required on one side. Sadly, we are dictated by builders and carpenters who have not a clue about brickwork.

The solution would have been to use a cutting system as shown in the diagrams above and left, where three-quarter bricks were used.

This would have allowed the perpends to remain aligned and would have given a much stronger bond.

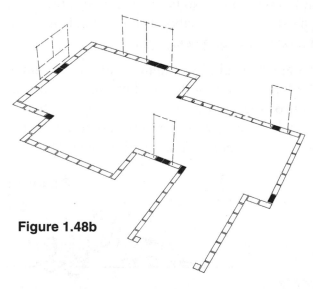

**Figure 1.48b**

This figure shows where the cuts can go to be least noticeable in difficult cutting circumstances. Internal corners under windows and doors especially where there is the least brickwork underneath or on top of the openings.

# 2
# Introductory Skills

## Holding the Trowel

Start by holding the trowel firmly so that it cannot be pulled from your palm, but so you still have fluency of movement. Most of the movements in bricklaying are either rounded or circular movements, for example, tipping of a brick or a block and the cutting off of excess mortar.

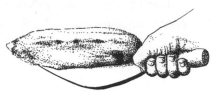

Figure 2.1

## Knocking Up a Mortar Board

The first stage in bricklaying is the art of knocking up a mortar board or working the mix to keep the mortar to the right consistency. This action is very hard to visualize by anyone who has not previously seen or had the use or experience of a trowel. I find that it helps beginners tremendously if they mix the mortar with their trowel from dry and then bring in the water to make a workable consistency. Because much more effort is required, experience is gained more quickly.

I find mixing mortar in a barrow with a shovel helps beginners get their wrists working in the natural rhythm required to mix mortar. No bricklayer enjoys the effort of knocking up mortar boards, so the most proficient method is understandably soon learned, and the useless movements in mixing mortar quickly omitted.

Using rounded and circular movements, begin by skimming the trowel along the bottom of the mortar board to the base of the mortar, picking up a trowel of mortar and virtually dumping it on top of the existing mortar, turning the trowel over, creating a tumbling effect. This is taking mortar from a lower point to a higher point and tipping it off. Do this to about ¾ of the mortar from the right side.

Figure 2.2

Figure 2.3

Now reverse the procedure from the left side of the mortar going ¾ of the distance back through the mortar. These motions are forehand and backhand.

**Figure 2.4**

**Figure 2.5**

Repeat this procedure, screening the mortar as shown. Any lumpy mortar can be shuffled to one side of the mortar board and broken down more easily with the back of the towel by pressing the lumps between the trowel and the mortar board.

**Figure 2.6**

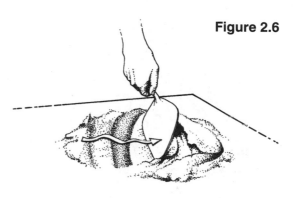

Forehand and backhand movements and rolling the mortar over are the way to mix fresh mortar with mortar already on the boards. This is also the way water is added to the mortar, if it is in need of freshening up to improve its workability.

**Figure 2.7**

**Figure 2.8**

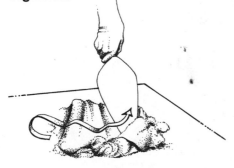

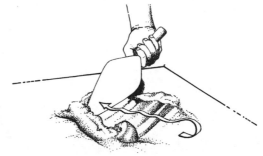

# Picking Up Mortar from the Board

Trowels of mortar can be filled anywhere from the mortar board. For keeping the mortar board clean and for better accessibility, the mortar should be troweled to one corner of the board.

Roll the mortar as shown using the trowel in a cutting motion.

**Figure 2.9**

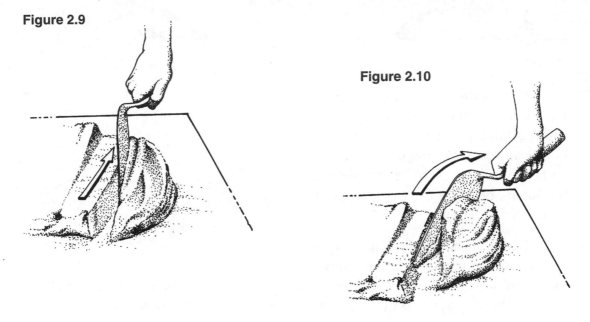

**Figure 2.10**

**Figure 2.11**

The trowel can be filled as shown.

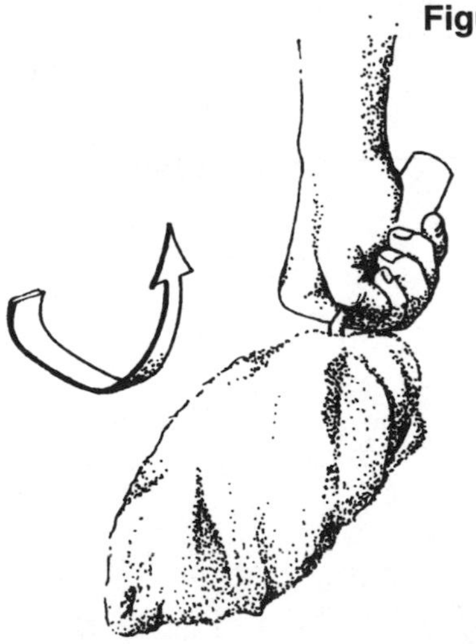

# Throwing Out a Mortar Bed

This can be practiced on a mortar board. Trowel your mortar to one corner of the board so you have access from one corner diagonally to the other. On an average mortar board of 2 ft (600 mm) square, this gives you room of 30" (850 mm) diagonally across the board, approximately 3.5 bricks.

**Figure 2.12**

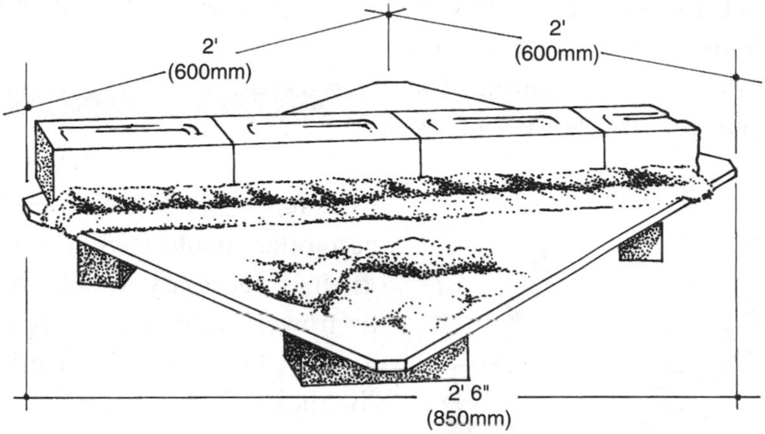

# Grasping the Brick

This I will describe for right-handed people, but if you are left-handed the same procedure applies, except vice versa.

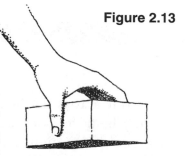

**Figure 2.13**

Grasp a brick in your left hand, placing four fingers on the left side of the brick and your thumb over the brick and on the right side as shown in the illustration. Now look carefully at the brick for any defects, chips, or flaws. In time, this inspection procedure will become second nature. As previously discussed, it is vitally important to save time and money.

# Buttering the Perpend

As you grasp the brick, the end pointing forwards is the front perpendicular face of the brick or front perpend. With the brick in the left hand and trowel in the right, pick up about a quarter of an average cup full of mortar. [The amount of mortar required will vary on the style and size of the bricks being buttered, but usually about a ¼"–⅜" (8–10 mm) thickness for a square brick and up to ½" for sandstock or tumble bricks.] Pivot the front of the trowel upwards so the mortar slides to the center of the trowel.

Now flick the trowel downwards about 2¼" (60 mm) quickly and bring it to a sharp halt. This will spread the mortar over the trowel. The test to see

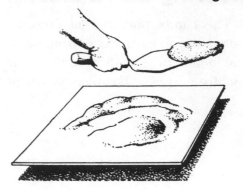

**Figure 2.14**

**Figure 2.15**

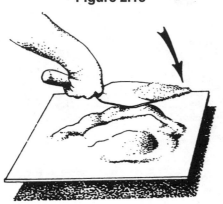

if this has worked is to invert the trowel and the mortar should stick to it without falling off. It is necessary to achieve this because the trowel—although rarely turned upside down in buttering—is inverted and tilted. The flick creates suction.

Now without moving the brick, and by using a slightly downward and circular motion, move the trowel down the front edge of the brick, pasting about ⁵⁄₁₆" (8 mm)

of mortar across the brick's end. By moving both the brick and the trowel in a circular motion, push the brick right and the trowel left. The brick is turned in towards the body and partly rotated clockwise as the trowel is wrapped around its face and is then curved counterclockwise.

**Figure 2.16**

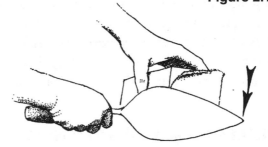

Moving the brick counter-clockwise and the trowel right, tip the right side of the brick.

This completes the buttering of the front perpend.

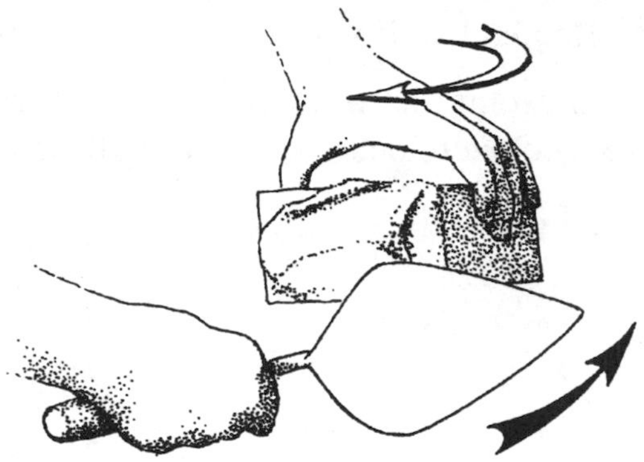

If the above steps appear a little difficult to grasp, then spend 10 min watching a tradesman brickie and you'll soon catch on. It's practice from

**Figure 2.17**

**Figure 2.18**

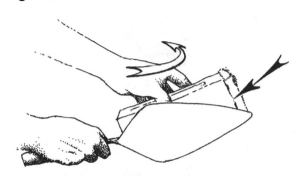

here on. You should always aim to complete the buttering of the brick in the three moves shown, as any additional handwork can lose you hours over the course of a day.

Both arrises of the perpend must be buttered sharp and full all the time so as not to have any problems with double-face brickwork.

The next step is to be sure to pick up mortar every time you pick up a brick. This is a little trick of the trade that so many people overlook. Simply, you should scoop up your mortar when you are bending to grasp the brick — otherwise it means double-bending for each brick picked up.

This may not seem too significant — but remember each standard house contains 7000 or so bricks — and that's bending down 14,000 times. It relates back to our golden rule and that is to obtain maximum efficiency with the least possible effort. I call it working smart as opposed to working hard. Success as a tradesman or contractor is about productivity and quality.

# Buttering the Back Perpend

Remember, if you are left-handed, just reverse this procedure.

Grasping the brick in the same manner as for the front perpend, rotate it 90 degrees counterclockwise exposing the back of the brick.

**Figure 2.19**

**Figure 2.21**

**Figure 2.20**

**Figure 2.22**

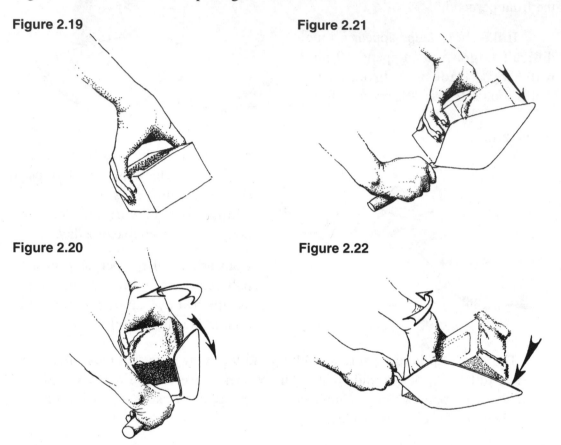

From here the same movements are used to butter the brick rear as those used for the front perpend.

# Laying Bricks

If you are right-handed then you lay bricks with the left side of your body against the wall working either backward or forward. Most brickies prefer to work backward. Pick up a trowel of mortar and follow the steps learned in "Throwing out a Mortar Bed" and "Buttering the Perpend."

To make sure that you maintain half bond, I recommend standing a brick up on its end to occupy half of the underneath stretcher. Be sure to keep the ⅜" (10 mm) away that is required for the mortar joint.

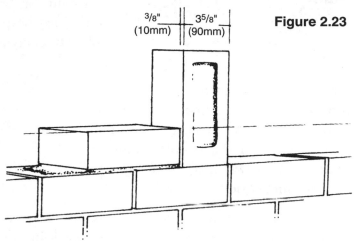

**Figure 2.23**

Place the brick on the mortar bed and slowly press it into position, keeping an even perpend joint.

**Figure 2.24**

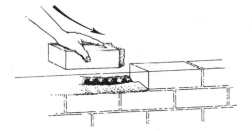

**Figure 2.25**

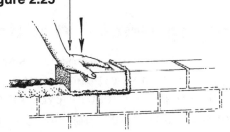

**Figure 2.26**

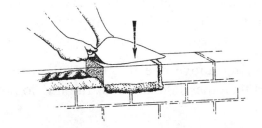

Make sure your thumb does not foul the stringline. Be sure to look down the face of the wall as you position the brick to ensure it remains flush to each face of the wall. Allow the trowel to true the brick as it cuts across the faces picking up the excess mortar. (Try not to drag your trowel across the brick faces because this will smear and embed the brickwork with mortar.)

**Figure 2.27**

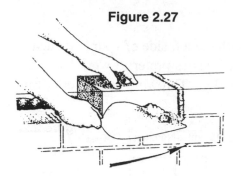

Using the edge of the trowel, tap the top of the brick to bring the brick exactly in line with the stringline. If the brick is tipped up or down at either end it will destroy the look of the wall as it will be easily spotted when surrounded by well-laid flat bricks. In the trade we refer to bricks that have been laid with ends tipped as "blind bricklaying."

**Figure 2.28**

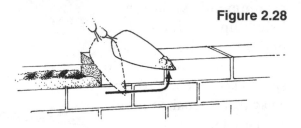

When cleaning off excess mortar always use the trowel in a cutting motion making sure the cutting edge of the trowel remains vertical. This will also make raking and cleaning easier.

It's my experience that there is no worse job than having to fill holes in perpends or gaps in mortar beds before jointing or raking. Bring the trowel through the bed joint and up the perpend joint, retaining the excess mortar to butter the next brick.

# Laying Bricks Working Forward

This process is the same as working backward. However, it is imperative to sight down the perpend joints to make sure they are aligned. Plumb them continuously vertical by eye. This point is very important and separates the professional from the amateurs.

To illustrate this point I ask you to imagine the consequences of two layers who start at either end and work in toward each other. The last brick in the center of the wall should fit straight in if both layers have kept their joints vertical. When laying single skin brickwork do not lay bricks over your easy eye height as it is difficult to keep the brick back on a level line and some tipping "off the plane" could result.

I've seen some jobs done where laying above eye height resulted in a "good" face wall with ⅜" (10 mm) beds but the other side had beds running to ⁵⁄₁₆" and ½" (8–13 mm).

**Figure 2.29**

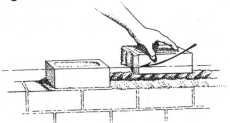

**Figure 2.30**

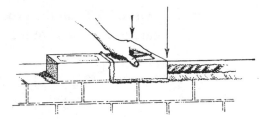

**Figure 2.31**

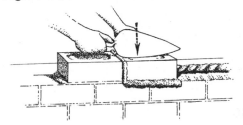

**Figure 2.32**

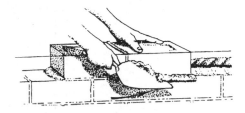

**Figure 2.33**

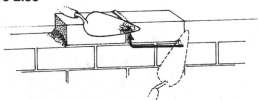

When laying bricks over the heads of windows and doors, especially if the brickwork is over your head, keep checking the wall face for plumb as the tendency is for the wall to run out and lean in towards you. This is a peculiarity of construction that is difficult to explain, but it is related to parallax error in sighting.

It is also equally difficult to keep the perpends plumb when laying bricks above eye level. This is because it is easier to sight down than it is to sight up, and when you are working close to a wall and above eye height it is only possible to sight upwards.

On houses with gables, it is best to build up each corner of the wall into the gable. Some brickies lay one brick at each end and then tingle or stretch the line up as they build up. This is acceptable, but be sure to use a long level and plumb down the face of the wall, otherwise the potential is there to have the top of the wall run out as described previously.

# Finding the Bond Vertically

Although we work from the ground up, we need to ascertain the high point to be reached to maintain the bond at the top as illustrated.

**Figure 2.34**

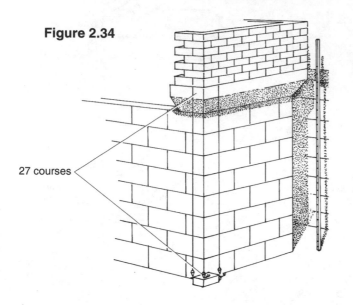

27 courses

This is undoubtedly one of the most difficult aspects to master in precision bricklaying, but with practice and patience a good tradesman will never have any difficulty in working out the right bond.

*Step 1.* Measure the distance from the top of the foundation to the underside of the first of the brick courses on top of the nib. Divide this distance by the appropriate gauge, which is found by adding the height of a brick to the mortar bed. Refer to workings and scale at the back of this book to determine the number of courses needed. The system below will give the correct starting bond.

**Figure 2.35**

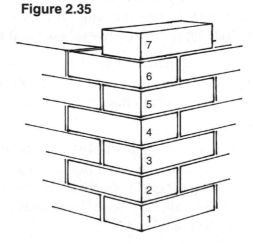

*Step 2.* Using a stretcher as the starting bond on the corner, we can see that every course that is an odd number (1, 3, 5, 7, etc.), will be a stretcher. Therefore if the first course on the corner is a stretcher, course 27 will start with a stretcher. (See Fig. 2.34.)

*Step 3.* Using the same bond as previously (starting with a stretcher on the corner), the bond on the second course will be a header. From this we can see the courses which are even numbers will be opposite bond, for example, 2, 4, 6, 8, 10, 12, will always be a header, in relation to the first course that started with a stretcher. If, for example, our figure had 26 courses, the corner should be started with a header course.

*Step 4.* Having worked out the correct bond for the first course, the brickwork can be plumbed down and gauge marked. A corner can now be erected.

**Figure 2.36**

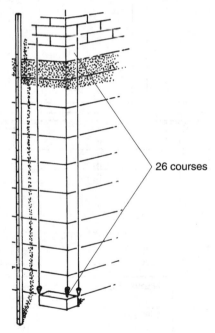

26 courses

# Gauging Brickwork for Length

The most common brick size is 7⅝" (190 mm) long × 2¼" (57 mm) high × 3⅝" (90 mm) wide. To the length of 7⅝", add ⅜" for a perpend, giving a total of 8".

This length is a good denominator for windows and doors of 2' (0.6m), 4' (1.2 m), 6' (1.8 m), and 8'(2.4 m) width as it makes it easy to work brick lengths into any openings that are in modules of 2'.

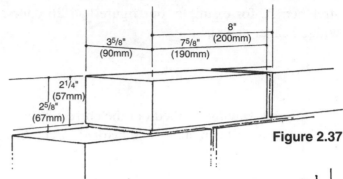

**Figure 2.37**

**Figure 2.38**

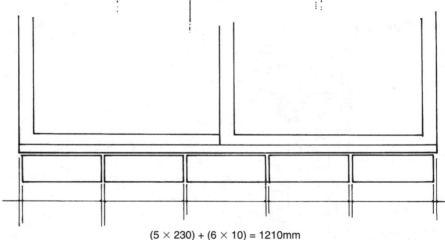

(5 × 230) + (6 × 10) = 1210mm
5 bricks + 6 perps = 1210mm

Use a Modular Spacing Rule
to set out.

| Name of Unit | Thickness | | Length | |
|---|---|---|---|---|
| | in. | mm | in. | mm |
| Standard | | | | |
| Modular | 4 | 100 | 8 | 200 |
| Engineer | 4 | 100 | 8 | 200 |
| Economy | 4 | 100 | 8 | 200 |
| Double | 4 | 100 | 12 | 300 |
| Roman | 4 | 100 | 12 | 300 |
| Norman | 4 | 100 | 12 | 300 |
| Norwegian | 4 | 100 | 12 | 300 |
| Utility | 4 | 100 | 12 | 300 |
| Triple | 4 | 100 | 12 | 300 |
| SCR | 6 | 150 | 12 | 300 |
| 6" Norwegian | 6 | 150 | 12 | 300 |
| 6" Jumbo | 6 | 150 | 12 | 300 |
| 8" Jumbo | 8 | 200 | 12 | 300 |

So think about this before you start opening or tightening perpends. It will eliminate a lot of cutting around windows. Even on low-set bases check the plan and try to "work" any cut under or to a window or door that doesn't "work" brickwork. For example, windows or doors of 3' or 5' openings don't easily accept brickwork. It works the same on high-set houses if the top windows are plumbed off the base, so give some advance thought to placing your cuts.

**Figure 2.39**

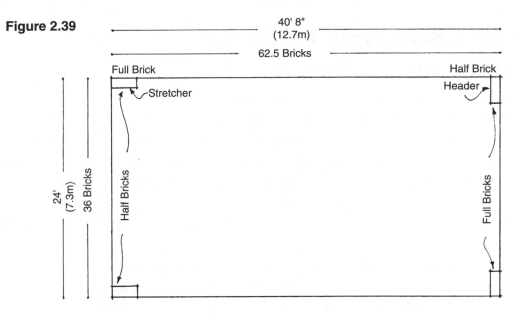

Now if we take a rectangular base, 40'8" (12.7 m) long by 24' (7.3 m) wide (these are common measurements for high-set bases) the following will apply:

If our brick work is to 8" length, dividing 40'8" by 8" gives 62.5 bricks, or 62 full bricks and one half brick. So the bond for the 40'8" wall will be a full brick to a half. (Header to stretcher). Write the bond on the plan.

Now take the 24' end: dividing it by 8" gives 30 full bricks. Our bond on the 24' end will be the same (either a stretcher to stretcher or header to header).

As you are laying the first course of bricks, measure 10 bricks with the buttered perpends; this is a good way to check and make sure it is the right length.

Bricks often vary in length and, when different types or colors of bricks are used on the one job, different lengths will occur. I use this solution which has been most helpful.

Take an average length by measuring 10 bricks, approximately 6'8" (2000 mm). If there is a blend of 50 percent one color and 50 percent another, my solution is to place two bricks together with a pencil joint between. Then measure the total distance allowing a pencil joint on the end thus allowing two perps. Divide the measurement by two and you have the module. This can also be used for fences, houses which are set on slabs and any areas where there is no brick base to follow. You can also apply this method for as many different bricks as are in the blend. Lay the bricks end to end with a pencil joint between and one at the end, measure the overall length, and divide by the number of bricks.

# Gauging Brickwork for Height

## Finding the Appropriate Gauge for Bricks

Measure seven bricks dry. This will equal six bricks and one brick for a bed joint. Substitute the right gauge for the right bricks, knowing that tumble and sandstock bricks will look better with a slightly larger mortar bed and perpend joint.

For example, the overall measurement of seven bricks is 15¾" (399 mm), but for sandstock or for the heritage range of bricks that are slightly bigger, it is wiser to stretch this measurement out to about 16" (400 mm) so the gauge will fit better. Brickwork gauge can be changed up to 1⁄16" (1 mm) per course without looking out of place, but anything greater than this will be quite noticeable.

Any changes to courses should be within the first 10 courses or above eye level height so it is least noticeable. Square bricks look better with a tighter gauge, and I therefore recommend that these are laid with a gauge of 15¾" (399 mm) as compared to our normal 16" (400 mm) gauge.

The four most common gauges for brickwork are 2⅝", 2¾", 3⅜", and 3½", so contractors in the industry buy a length of aluminum 7'8" (2.3 m) long and 1" (25 mm) square and mark on it the four common gauges.

**Figure 2.40**

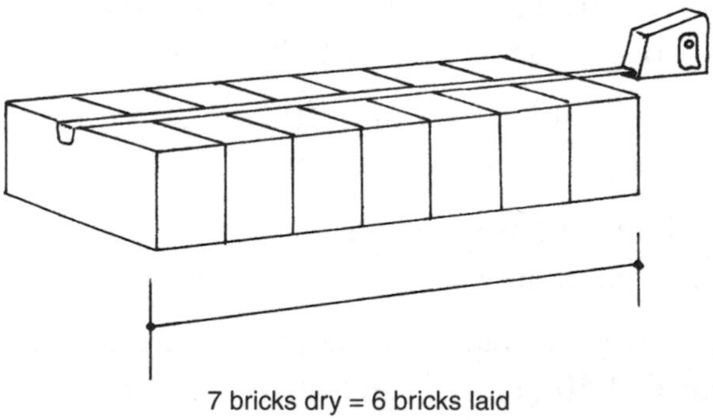


| Name of Unit | Height in. | Height mm | Modular coursing | Modular mm |
|---|---|---|---|---|
| Standard | | | | |
| Modular | 2⅔ | 67 | 3C = 8" | 200 |
| Engineer | 3⅕ | 80 | 5C = 16" | 400 |
| Economy | 4 | 100 | 1C = 4" | 100 |
| Double | 5⅕ | 135 | 3C = 16" | 400 |
| Roman | 2 | 50 | 2C = 4" | 100 |
| Norman | 2⅔ | 67 | 3C = 8" | 200 |
| Norwegian | 3⅕ | 80 | 5C = 16" | 400 |
| Utility | 4 | 100 | 1C = 4" | 100 |
| Triple | 5⅓ | 135 | 3C = 16" | 400 |
| SCR | 2⅔ | 67 | 3C = 8" | 200 |
| 6" Norwegian | 3⅕ | 80 | 5C = 16" | 400 |
| 6" Jumbo | 4 | 100 | 1C = 4" | 100 |
| 8" Jumbo | 4 | 100 | 1C = 4" | 100 |

Be sure to mark gauge on each side of the gauge rod, and then mark the head and sill heights for windows and doors. Having done this, you have a gauge to allow the precutting of all headers and sills for doors and windows.

When measuring for headers under the windows on walls constructed only of single skin brickwork, be sure to deduct ⅟₁₆" (2 mm) to allow for the thickness of the rubber weatherstrip fitted to the underside of the window.

Most windows are made in modules of 12" (300 mm), both for their height and their width. This is good for bricklayers because the gauge can be penciled on the gauge rod at 2', 3', 4', 5', 6', and 7' (0.6, 0.9, 1.2, 1.5, 1.8, and 2.1 m).

Talk to the carpenter after the foundation brickwork has been completed and inform him or her of the gauge so he or she will know at what height to install windows and doors.

Communication with fellow trades-people makes every-one's job just that little bit easier.

There is nothing to be lost by making a policy early in your career to have discus-sions where needed with other contractors or tradespeople work-ing on the same job to set out the require-ments to efficiently assist each other in your tasks.

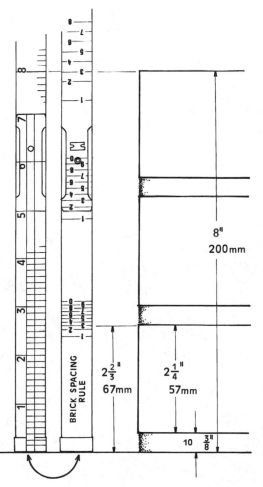

Using the spacing rule — Check the plan for the number of courses to correct height. To deter-mine correct gauge the plan should give specific measurements eg. 3 courses to 8" or 200mm. Locate 200mm on the spacing rule, turn the rule around and it will tell you what course to use. In this case 3 is the correct gauge for 200mm (8") or 6 for 400mm (16").

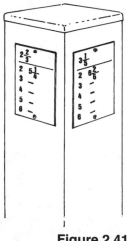

**Figure 2.41**

# 3
# Setting Out the Job

## Introduction

The figure below shows a typical set up for one of my jobs, I consider the most important factor is programming, and also remember to make sure the other trades have completed their work. There is no sense turning up with eight bricklayers and

**Figure 3.1**

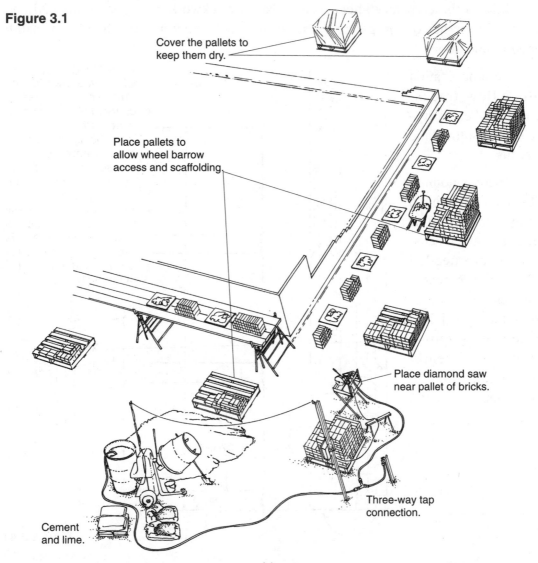

Cover the pallets to keep them dry.

Place pallets to allow wheel barrow access and scaffolding.

Place diamond saw near pallet of bricks.

Three-way tap connection.

Cement and lime.

four assistants when the house hasn't been prewired or the plumber hasn't completed the "rough-in." Next, ensure the materials are in their right places—this could mean in some cases having the bricks delivered before the footings are poured, as accessibility could be impossible later when construction is underway or the slab is down. Being on site when the bricks are delivered and having a knowledge of the construction helps as the pallets can be placed as required.

If the construction has a series of full-length windows on one side only place enough bricks on that side to cover that wall. I like to position the pallet loads of bricks about 8' (2.4 m) from the wall. This gives enough room to erect the scaffold and to wheel a barrow around.

Mobile phones have made life so much easier. They make it simpler to organize most things quickly and I like to have the cement lime and loam delivered at about the same time. The cement and lime can be stacked on a pallet and covered with plastic beside the loam. I try to put a brick-cutting saw beside one of the pallets to save time carrying bricks to the saw. Notice the hose in the illustration? Get yourself a three-way tap from any hardware store. It enables the saw to be working and the drum and wash buckets to be filled without any hassles. The hose in the drum also has a nozzle that can be turned off or left dripping to keep the drum topped up.

Secure all electric leads up out of the way. The lead that goes to the saw is placed up on a saw stool to keep it free from water. Always, always use a grounding unit. Water and electricity do not make good companions on a building site.

## Construction Site Set Out

The two most important points to setting up any job are:

1. Run lines to profiles and check all measurements first.

2. Take a level on all corners on the site.

Check the footing for level, and find the highest part of the concrete footing. Use this as the reference point, as it is easier to bed up a little on the first course of bricks around the base of the foundation than to grind down your gauge. Levels are easier to rectify from the start and should be correct before coming above ground level.

Be careful of this situation—a parallelogram, where all sides could be equal, but end up out of square. Always run a tape over the diagonals and shorten the longer side or lengthen the short side. The dotted line is true square.

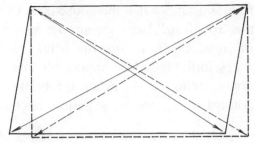

**Figure 3.2**

# Corners

Corners are built first. The usual practice is to allow one person to erect the corner from the bottom to the top, and it is his responsibility to keep the work plumb and level and to the correct gauge.

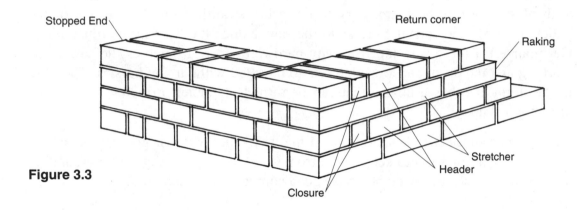

**Figure 3.3**

When building up corners using profiles, too much adjustment and fiddling around can make the corner "floppy." It's best to get in and build the corner without too much tapping and so on.

This has become more apparent in recent times with the advent of profiles that have made a brickie's life easier but have somewhat reduced the skill required in building corners.

The important point is to think about what you are doing. Having the profiles in place means a corner can be constructed very quickly, but if you are ⅜" (10 mm) out when you reach window height you have defeated the purpose and the advantage of using profiles.

## Setting Out a Corner

This involves plumbing down off stringlines as in the following illustrations (most commonly), or measuring off existing concrete formwork.

**Figure 3.4**

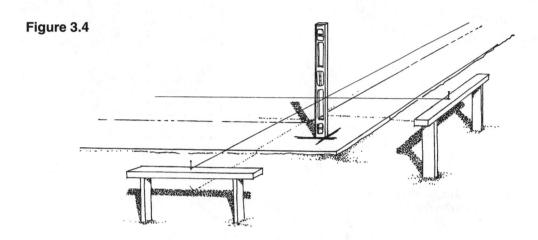

**Figure 3.5**

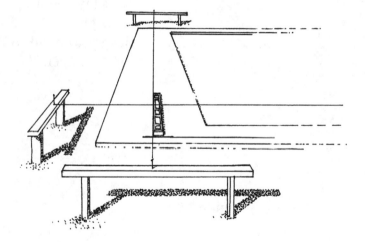

## Where to Begin the Brickwork

After establishing your height, go to the corner that is the lowest part of the site and work upwards from there. Most corners are 90 degrees, irrespective of which way you start. Now remember our reference point is the highest point of the slab or footing and it is from this we determine our datum point.

After verifying a gauge by the system shown in the figures given at the back of this book, mark your gauge down from the datum point.

**Figure 3.6**

A datum point is six courses above the reference point.

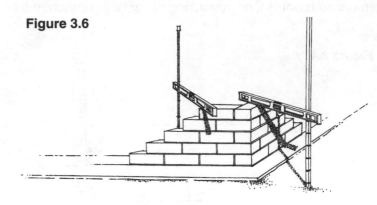

**Figure 3.7**

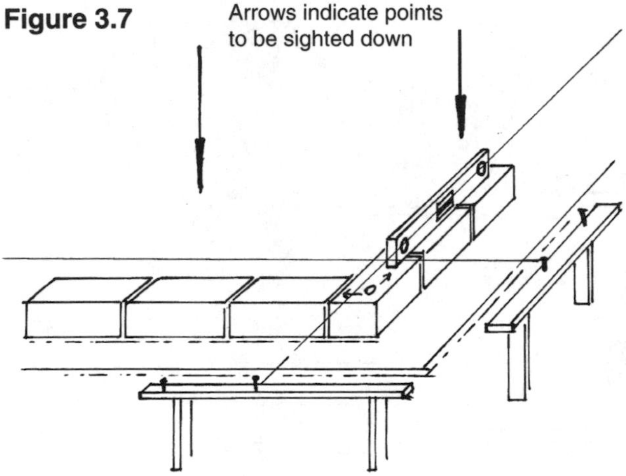

Arrows indicate points to be sighted down

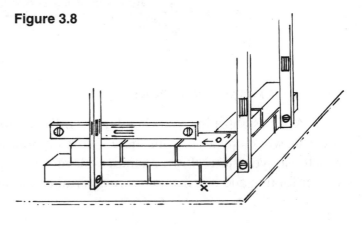

In this case, butter three bricks and lay them up to the header. This gives $3 \times 3.5$ bricks, enough length to build to the full height of the corner.

**Figure 3.8**

Now carefully continue sighting vertically down the corner after each course is complete.

*x* indicates points on corner to be plumbed;

*o* indicates points on corner to be leveled.

Plumb the corner, making sure the bricks sight parallel with the line, and level every course. If there is no peg or steel in the corner to mark and check the gauge on, measure it.

**Figure 3.9**

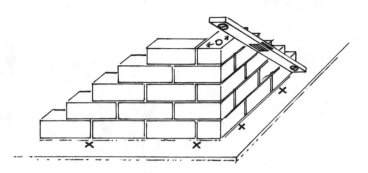

When the corner is complete, profiles can be fixed, and the brickwork continued to full height.

It is the responsibility of the person building the corner to constantly check that the gauge will work to windows and door head heights, and where necessary, sight the tingle through.

# Internal Corners

Internal corners can be visualized as building an external corner from the inside. It is important to keep check on the back of the internal corner, to make sure that all bricks are laid flush. This is done by sighting the two faces as shown.

**Figure 3.10**

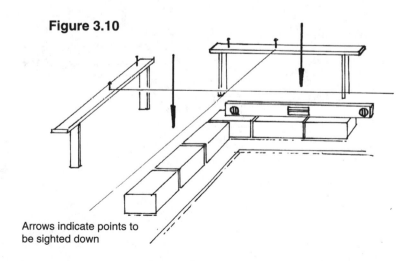

Arrows indicate points to
be sighted down

**Figure 3.11**

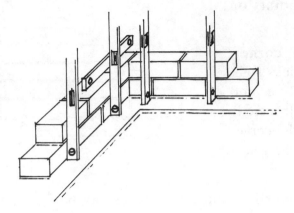

The diagrams on this page indicate points to be plumbed and leveled.

**Figure 3.12**

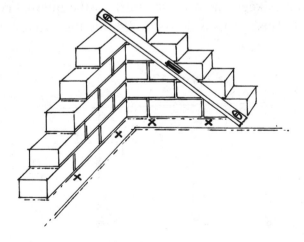

# Moving Up in Stepping Foundations

This is probably one of the hardest tasks to accomplish after building the corner.

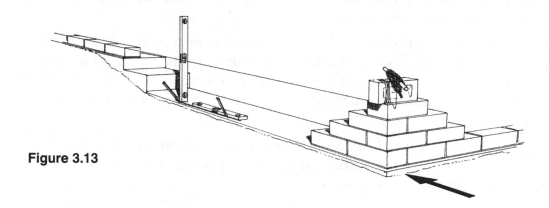

**Figure 3.13**

First, move along to the footing at the opposite corner. Determine the appropriate bond as shown in the system discussed earlier, and place one brick to the right gauge and bond. Run a stringline from this brick across to the brick on the opposite course.

Now move to the lowest end to the first step. Plumb down directly from the stringline at this point and mark a parallel line approximately 1'8" (500 mm) long by laying the level down and sighting it through as shown. Measure from the line to the footing at this point to establish the height of the first brick. The thickness of mortar may need to be varied depending on the height of the footing. Having done this, block the line down to the first course and continue to lay the bricks from the established corner end.

The rest of the footing can be done the same way, but remember to keep the brickwork half bond. This can be checked by placing a half brick directly on top of the brick underneath and in line with the perpend—which should give you a ⅜" (10 mm) per-pend over. Keeping the brickwork on bond assures strength, as well as a neat appearance.

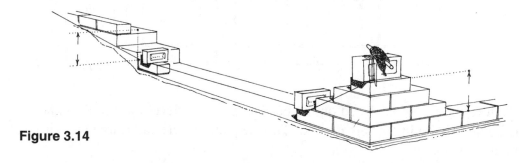

**Figure 3.14**

# Working Out Window and Door Heights

Nearly all houses maintain window head height to suit the door height. Windows and doors will look much better with stretcher bond over them, rather than upright soldier courses which always seem out of place.

On most high-set brick house bases, the steel door frames and the aluminum sliding doors and windows can be adjusted up or down by the bricklayer to maintain gauge if the concrete floor hasn't been poured. The concrete floor can then be poured to suit. The standard size height of doors is 7' (2.1 mm). This works 84 gauge or 87.5 gauge.

Even if you were lucky enough to have brick work 84 gauge or 87.5 gauge, most of the bricks have been extruded (holes) and won't be covered by the door frames. This is also true with windows.

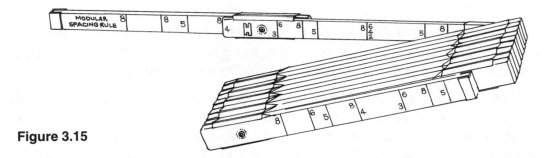

**Figure 3.15**

The only way to hide them is to turn the ends up into a header course under the doors and windows. First work out the minimum floor height—usually this is about 1' (300 mm) out of ground at the highest point. After working out the appropriate gauge, run the brickwork up *x* courses making sure to leave out where the doorway openings are. Cut and lay the headers to the appropriate size.

**Figure 3.16**

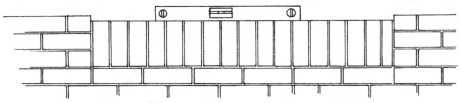

Before the frame is placed on the headers put your level on top of the headers and make any fine adjustments needed. This helps to ensure the frame will fit level, plumb, and square.

Stand the frames on the headers making sure to parallel the frame by measuring each end of the jamb off the line.

**Figure 3.17**

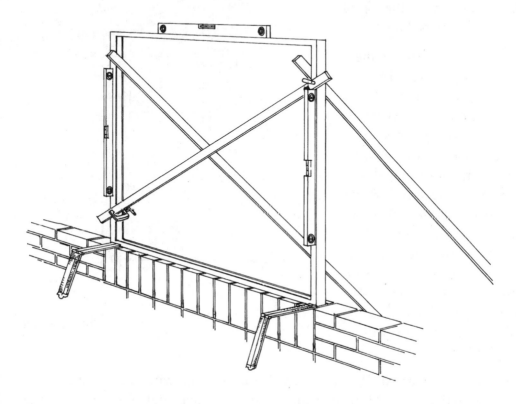

Next, fix stays and plumb the frames.

**Figure 3.18**

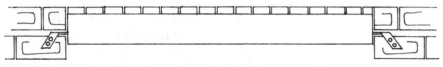

Plan of window showing fitting of window ties into cavity.

# Brickwork and Meeting up with Soffits

Depending on the preferred type of construction, some builders prefer that the soffit be level with the top of the windows in the house.

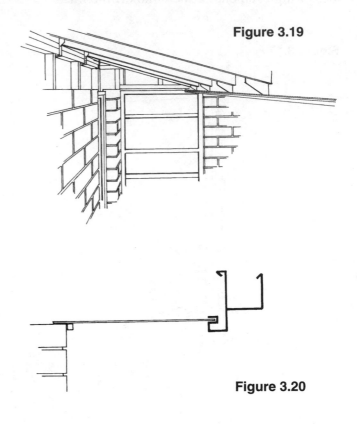

**Figure 3.19**

Almost everyone these days on commercial and domestic jobs is using fascia that has a rebate to accept the edge of ¼" (6 mm) versilux, to be used as a soffit (Fig. 3.19). To make the soffit level with the top of the window, we run our bricks up either side of the windows finishing ⅛" (3 mm) below the height of the windows. This allows the carpenter to fit the soffits and fasten a ¾" × ⁵⁄₁₆" (20 × 8 mm) molding to finish off (Fig. 3.20).

**Figure 3.20**

On jobs where the brick work will run past the window height and the soffit has to be nailed up to a pine batten, take care that your horizontal mortar joint around where the soffit will meet the wall will not be level with the soffit edge, otherwise the molding will fit into the mortar bed and be difficult for the carpenter to secure (Fig. 3.21). I also find in our work that the carpenters are very grateful for our thought and efforts on one other small area to help ease the fitting of soffits. We lay an extra brick on top of our wall at each rafter. This allow the carpenter to easily fix a 2"×1" (50 × 25 mm) piece of pine down plumb from the rafter onto which the batten that the soffit will be nailed to can be fixed.

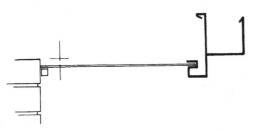

**Figure 3.21**

# Window Heights

When starting on a brick veneer house, check that all windows are the same level. This is important when working out your gauge. In some cases an area of the house may be stepped, either higher or lower to allow for a garage, a sunken living room, or a split-level design. This will mean a difference in height of some windows.

To avoid having Soldier courses over all the other windows, make an appropriate decision on the gauge, based on the height of the majority of windows. The use of "straight-edges" fixed to the outside of the house simplifies the brick veneering of a house.

**Figure 3.22**

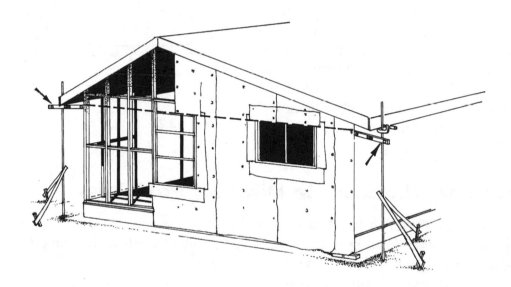

The posts are set and plumbed the required distance to the outside of the brick skin. Window head heights can be leveled onto the posts and the appropriate gauge marked down. A stringline can then be connected to the posts and the brickwork begun.

Sometimes windows or doorways in a wall will fall in an almost impossible position. Normally windows around a construction give the bricklayer the gauge for the job.

# Brickwork in Difficult Situations

Figure 3.23 shows a problem door frame. Clearly the frame has been fitted without much care and there is no way the brickwork will run up to it. The door frame cannot be shifted due to an internal wall frame, so start the brickwork on the other side of the frame allowing the same distance away as it is short on the offending side. In this way joinery can now be fitted on both sides of the frame finishing off the job professionally. With the introduction of premade wall frames this can occur quite often, especially on openings such as garages and internal or external corners where the brickwork returns to the openings. This method can also work horizontally. If you find there is a door jamb, either timber or aluminum, perhaps ⁹⁄₁₆" (15 mm) lower than all the windows, run either a piece of angle or quad timber to cover the gap between the door and the brick work. Placing a header over one door or window would look completely out of place and in brickwork, symmetry is very important.

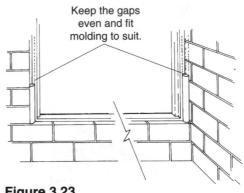

Figure 3.23

# Gauging Windows in Difficult Positions

Builders prefer to keep all window heads at the same height, but where a carport is stepped down or a kitchen window is fitted flush with countertop, the alignment of the top of the opening with all adjacent openings becomes impossible (unless the doors or windows are custom made).

In these cases soldier bricking is used across the head of the window or doorway to make up the coursing.

When you have the job of fitting a window, work out your courses and gauge down from the top of the window.

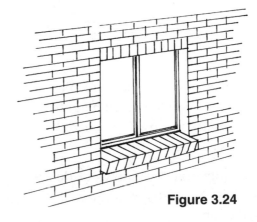

Figure 3.24

That is, make sure that the brick course over the windows will not have to include soldiers or cut brick. Measure the height of the window, fit the sill, and then brick up to either side.

**Figure 3.25**

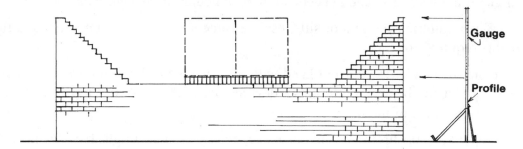

In Fig. 3.26 the bricklayer fitted the window at the same height as the door and used soldiers top and bottom. Together with the headers over the doorway, this job shows a lack of planning and professionalism.

**Figure 3.26**

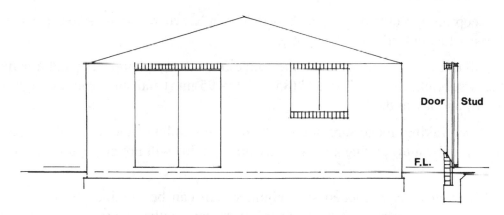

Disregard door heights in your planning—they rarely work brickwork. Use the windows in your construction to determine your course and plan ahead for the odd window out.

# Window and Door Sills

With all headers and soldier courses above and below windows and doors, buttering headers and sill bricks is done the same way.

The difference between a header and a sill brick is that the sill brick is laid at an angle to the wall face, and headers are perpendicular to the wall face.

When buttering headers or sills, use the same amount of mortar. Almost a full trowel is required for each.

Hold the header in your left hand with the cut edge or bottom towards you and tilt the face of the brick slightly up 45 degrees. Paste the complete top of the brick giving a solid mass of mortar.

Depending on broken bricks or extruded bricks, you might have to pick up extra mortar to complete the buttering.

Using the trowel in a downward motion, trowel a bead of mortar on the two sides and the top of the brick in an even consistency — usually about ⅜" (10 mm) or ⁹⁄₁₆" (15 mm) thick all around.

There is no need to butter the bottom as the bed joint will squeeze up into the perpend joint. Laying one brick on either end of the sill first allows the use of a line between a pin at either end to provide a straight and even line to follow with the tops of the headers.

Keep an eye out for wedge-shaped perpends, where the mortar gap varies in thickness causing bricks to tilt slightly.

I find it is easier to mark out the gauge from the left-hand side; add a perpend for the first brick; 2¼" + ⅜" = 2⅝" (85 + 10 = 95 mm) and start from the right side of the sill and lay to the left.

When raking, make sure not to rake out the middle of the top of the header—if it is 3⅞" (110 mm) or single skin brickwork—as this will not be weather or vermin proof.

When you have reached sill height, which can be obtained by use of your gauge we spoke of earlier, mark the opening and lay another course of bricks leaving the brickwork out where the windows fit. Now lay the headers, and after completion sight along their top edge.

Put your level across the length of the headers in the middle of the opening and make sure all the headers are straight and level. This is where the window will stand so it must be right.

Measure in 1⅟₁₆" (27 mm) from the outside face of the header bricks at each end, and draw a pencil line across to show where the window or door will fit.

A good idea is to spread a thin bed of mortar in the middle of the sill. Sit the window on the mortar bed on the headers to the pencil line, and gently tap down the frame into position.

**Figure 3.27**

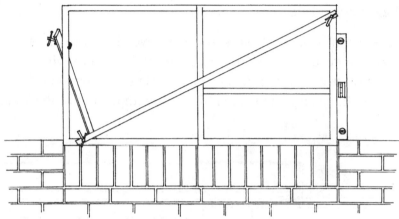

Plumb and check for level, and square it up while bracing the window. (I personally have manufactured a set of window props which are hammered into the ground and fixed to the window with self-adjusting sliding hooks. I've seen pieces of 2" × 1" pine used equally as effectively utilizing some sliding or G-clamps. It doesn't matter how you do it, but that frame must be held firmly square and level.)

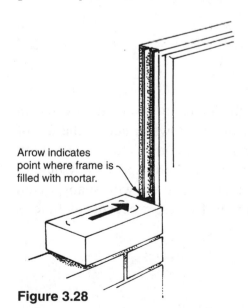

Arrow indicates point where frame is filled with mortar.

**Figure 3.28**

Check the height of the window for gauge, as you might have to adjust the coursing height several sixteenths of an inch. Check it as you lay up to the window.

Fill the reveals on the windows with mortar. This makes them weatherproof and stops daylight showing through.

Also attach the correct ties at the right intervals, depending on the height of the windows. The wire ties hold the window into the brickwork and are very important. On big windows, try to put in a few extra ties.

When brickwork reaches the window head, place your lintel bar over the window and align the bar so it is parallel at equal distance off the brick face. If the bars are a thin gauge, you might need to place a bed joint along the lintel bar to ensure bricks laid on it will still meet the stringline and correspond with the bonds and the courses at either side of the window.

In the event that your brickwork finishes several sixteenths of an inch below the window for whatever reason (and who can ever be sure what the reason is?), place a dab of mortar on the top of the brick close to the window frame. This will allow you to raise the lintel up to make it fit flush with the top of the window.

**Figure 3.29**

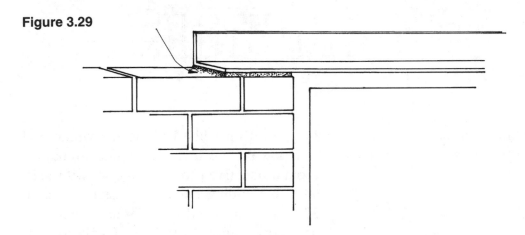

Lintel bars should always be propped with timber to eliminate any sag in the bar. This is particularly important on very wide windows or across the top of wide sliding glass doors, or where there are a number of courses to be laid on top of the window placing great weight down on the bar. I like to spray the windows with a penetrating oil or a vegetable oil that can be washed off with soapy water. This prevents burning or scouring of the aluminum frames, especially if lime is used in the mortar.

# Tricks of the Trade

To remain competitive in our changing work environment requires innovation on the part of the bricklayer.

**Figure 3.30**

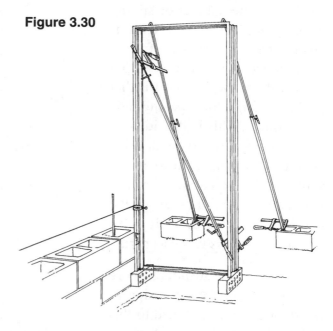

I developed my own diagonal brace for steel door frames because I couldn't purchase one commercially. My brace is shown in Fig. 3.30. It is made up of two G-clamps and two pieces of steel tube, one of which slides inside the other with a locking bolt, making the brace adjustable. It's light and portable, not to mention time-saving. At the top of the brace there is a piece of threaded rod that can wind in or out to give a pushing or pulling motion enabling door frames to be manually plumbed. This method is helpful for door frames slightly out of square.

# Making Sure Doors Are Plumb, Level, and Square

Fit the brace as shown in Fig. 3.31. This means turning the G-clamps around so the clamp fits on top of the door frame. The door frame can now be winched back into place. This method can also be used on windows out of square. A brace such as this can also double as a spreader in the middle of a sliding door frame

**Figure 3.31**

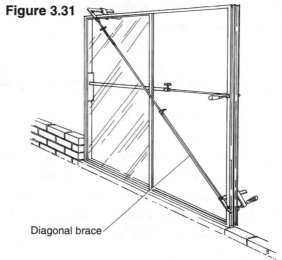

Diagonal brace

as shown in Fig. 3.32. And while we are on the subject of sliding glass door frames, I find it always pays to prop the frames with spreaders or at least to put the level up the brickwork that pushes against the opening side of the frame. This is the side that tends to bow inwards with brick pressure pushing against it. Be sure that it is square and level as it is hard to fix when all of the brickwork is laid. If it is in double brick, it is twice as hard.

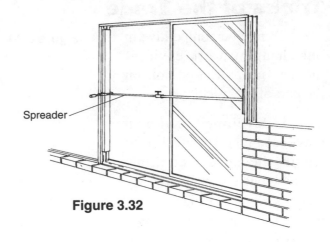

**Figure 3.32**

I also like to attach more ties than normal on this side as the door catch is attached to this side of the frame. Severe pulling of the latch could dislodge this side of the frame from the brick. The same care should be taken in a brick veneer situation, as bowing of the door jamb will make the fitting of security doors difficult or in some cases impossible.

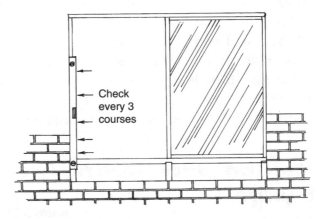

**Figure 3.33**

# Laying of the Sill (Brick Veneer Sills)

When laying brickwork up to and around a window in a brick veneer house, a damp course is needed under the course below the sill. If a course of split bricks is required under a window frame, then the damp course is run under this. This damp proof course, usually plastic or aluminum, prevents water from penetrating down the back of the sill bricks and onto the timber frame inside the wall. In both cases the damp course should be folded or turned into the nearest perpend past the outside of the window frame (Fig. 3.34).

**Figure 3.34**

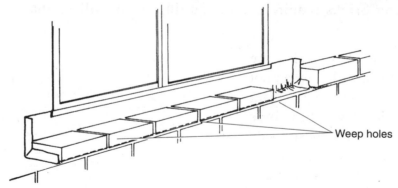

Weep holes

To prevent any water running into the cavity, leave weep holes to allow the water to escape (as shown above).

**Figure 3.35**

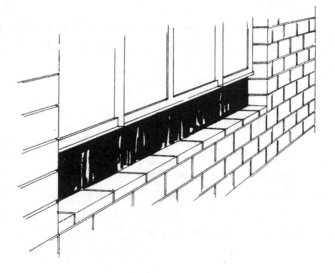

Laying splits under the sill will correct the steepness of the sill where necessary.

Most window sills require you to cut ⅜" (10 mm)-square pieces from the top corner of each end brick on the sill so the brick will sit flush under the rubber and not kick down at the top. The cut is taken out on the side nearest the window reveal. This is because of the manufacture of the windows and if care is not taken to do this the brick on each end will be lower, giving an incorrect line. If this happens, then the remaining sill bricks will be higher to meet the base of the window frame, or alternatively, if the remaining sill bricks are laid in accordance with a line from an uncut brick at either end of the sill, then a gap will eventuate under the window.

All frog bricks require reverse cutting of the sill so that frogs are not exposed.

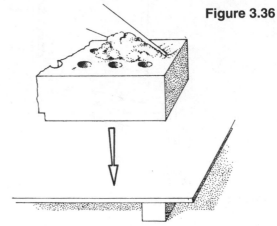

**Figure 3.36**

With extruded bricks, the holes have to be filled so as not to be obvious when looking across the window sill. This is best done from the opposite side of the brick so it is easier to clean the brick of excess mortar.

The best way to do this is to place a dab of mortar on the opposite side of the brick, hold the brick about 2⅜" (60 mm) above a flat level surface and bang the brick squarely onto the surface forcing the mortar through. This will make sure the mortar stays in the holes, but be careful you don't chip the brick.

# Buttering of Sill Bricks

Grasp the brick the same way as you would when laying bricks. With a full trowel of mortar, lift the bottom of the brick upwards to 45 degrees. Moving the trowel in a circular downward motion, paste the complete area with a solid mass of mortar. Now run the leftover mortar on the trowel around the perimeter of the brick, making sure to spread an even bead of mortar.

**Figure 3.37**

**Figure 3.38**

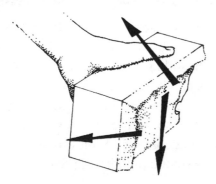

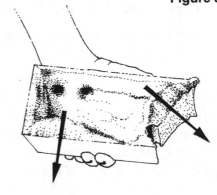

Place the brick on a bed of mortar on the sill and gently tap the brick into position allowing ⅜" (10 mm) of bed under the sill.

**Figure 3.39**

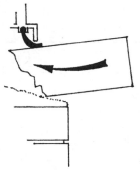

**Figure 3.40**

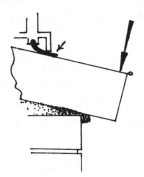

Parallel the sill by measuring as below.

**Figure 3.41**

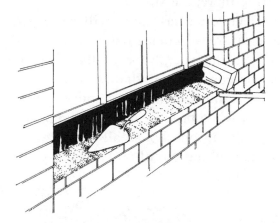

Go to the other end of the sill, or any other sill at that level, and repeat the process.

**Figure 3.42**

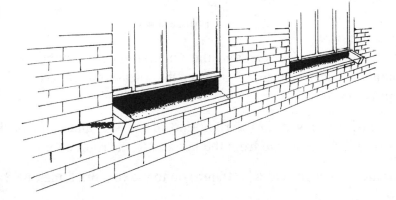

Lay all the sills on the one level with one stringline. If the wall is very long, place a tingle on the middle sill. Try to keep all the sills on the one level. Remember, however, that some sills may require splits underneath to correct the steepness of the angle, and to keep all the sills on the same angle.

While on the subject of sills, let's revert to brick sills. This is how I would construct my own house, but the time and the cost to do it this way make it totally impractical for commercial construction.

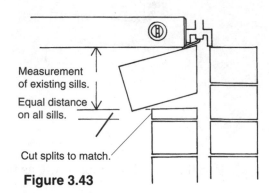

Measurement of existing sills.

Equal distance on all sills.

Cut splits to match.

**Figure 3.43**

Measure down from the underside of the rubber underneath the window frame (Fig. 3.43) to the top of the last course. If this measurement doesn't come to 4½" (115 mm) or more, the sill will be to flat to be effective and a "split" will be required underneath the sill brick. I find it best to measure the course below. If in doubt you need at least 7½" (191 mm). Now let's get technical. Check out the brick as in Fig. 3.44. It shows the bricks cut out on the edge of the window opening. Splits are then cut and fitted in between. It looks much better but requires much more time and effort and has to be allowed for when pricing jobs. There may be as many as twenty sills in one house all requiring "splits" cut to suit and the cutting out of 40 bricks or so could certainly blow the budget.

For neatness all windows in the house should have the same angle of sill. To attain this, I measure the difference

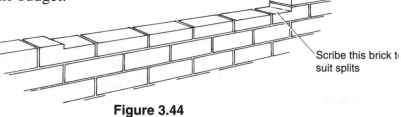

Scribe this brick to suit splits

**Figure 3.44**

between the last course of bricks and the underside of the window frame on the majority of sills of that height, then I transfer that measurement to the other windows and cut the splits to suit.

I find it easier to lay from the right end of the sill, and work back to the left end. If this is so, mark the gauge from the left end to the right end.

When marking gauge, add ⅜" (10 mm) on to the first brick making 3¼" (77 mm).

Pull the rubber on the bottom of the window so it is equal distance from each end of the window, and tuck the excess into the cross joint.

**Figure 3.45**

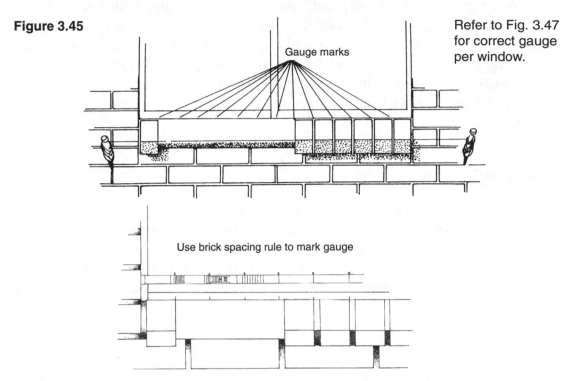

Gauge marks

Refer to Fig. 3.47 for correct gauge per window.

Use brick spacing rule to mark gauge

After the sill is completed, stand to one end and sight it through. If for any reason the bricks are bent, don't attempt to twist the bricks to try to bring them into alignment as this will be more noticeable than a slight wedge in the mortar joint or crossjoint.

**Figure 3.46**

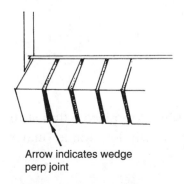

Arrow indicates wedge perp joint

When laying sills it is important to lay the bricks on the sill with a full perpend joint. The reason here is because they are the bricks most subjected to weather and running water on construction.

You will quickly discover when building sills that it is imperative to have ample mortar on the perpends which abut the framework of the construction.

This is because there is nothing worse than trying to joint up or rake sills with holes in the perpends. The mortar continues to fall in, and there will be a continual problem keeping out weather and vermin.

# Gauging Difficult Sills

When gauging for 3' (900 mm) and 5' (1500 mm) sills, which are difficult, first look at the gauge on the brickwork on the construction. If the gauge is 2⅔", (one brick plus a thin mortar joint), we would use ¼" joints or tighten the joints and cut one or two bricks in the center of the sill. If we were using ½" joints, which is a bigger gauge for a bigger brick. We can cheat a little by opening the ends or reducing the end joints (the ones against the window), as these are the least noticeable.

# Justifying a Gauge

Work 2⅝" (67) gauge for all windows in modules of 2'. For 3' or 5' windows which won't work 67 gauge, either open or close the perpends by making the mortar joints wider or thinner, or place one or more cuts in the sill bricks as required. The method is determined by the size or structure of bricks used. To lay the last sill brick, butter both sides of the brick and also mortar up the brick on either side of the final sill brick.

**Figure 3.47**

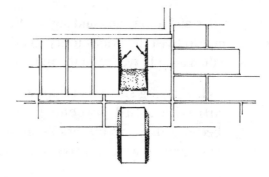

On houses with timber floors or a tiled roof, it is essential to place a thin strip of lattice under the window to fit between the bottom of the window frame (under the rubber weather strip) and the top of the brick sill.

**Figure 3.48**

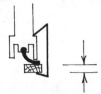

End elevation of window frame shows piece of lattice. A strip of expansion jointex can be used instead of timber. It can be left under the window to vermin- and weatherproof.

**Figure 3.49**

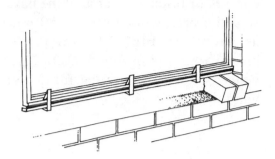

Make sure the rubber is even each end, and remove the lattice after completion of the sill to allow raking finish. If this is not done, the bearers and joints in timber houses could shrink up to ¾" (20 mm), compressing the window down onto the sill, bending the frame and making it difficult to open or close the window. Worse still it could crack the glass. The weight of a tiled roof will have the same effect. It's a point that's vitally important to remember.

## Floor to Ceiling Windows

On all jobs, start from the footings. A careful look at the elevations on the plan can save a lot of headaches later. Full-length "low light windows" as in Fig. 3.50 that sit on top of the concrete floor and require a brick sill underneath, have to be well planned.

**Figure 3.50**

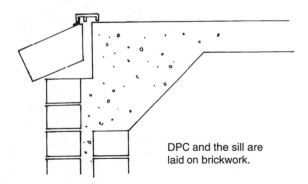

DPC and the sill are laid on brickwork.

When laying the top two courses of brickwork around the base, step the brickwork over 2" (50 mm) as shown in Fig. 3.50 for cavity brickwork or as in Fig. 3.52 for brick veneer, exactly where the full-length windows go.

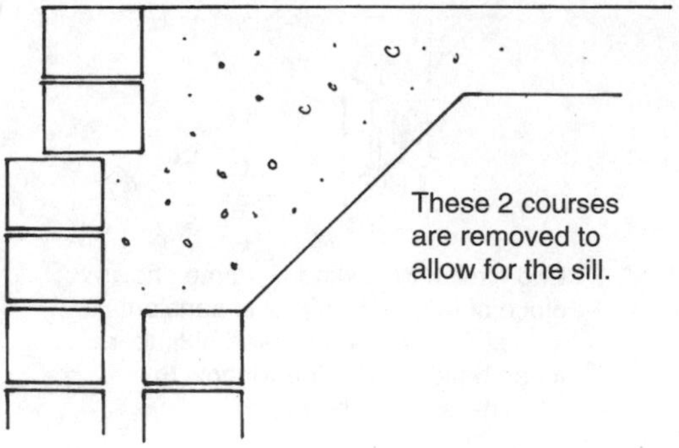

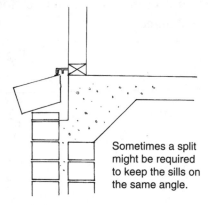

These 2 courses are removed to allow for the sill.

**Figure 351**

**Figure 3.52**

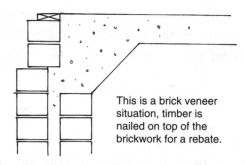

This is a brick veneer situation, timber is nailed on top of the brickwork for a rebate.

This will allow you to knock off the two courses of brickwork and replace them with proper sill bricks and splits.

**Figure 3.53**

Sometimes a split might be required to keep the sills on the same angle.

# 90° Return Corner Sill

1. Lay a pattern brick in correct position (Fig. 3.54).

2. Draw plumb lines 1-2 and 3-4 and level line 5-6.

3. Place three good cutting bricks on a flat surface.

4. Measure and mark off top slant length 5-1 across the face of the three bricks.

5. Measure and mark off horizontal length 5-6 both front and back.

6. Measure and mark off the bottom slant length 4-2 on the rear face of the bricks.

7. Mark horizontal length 5-3 on rear face and join 3-2 for rear cutting line.

8. Repeat the process in opposite direction for opposite miter cuts.

**Figure 3.54**

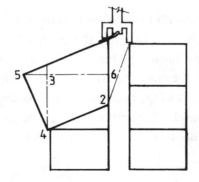

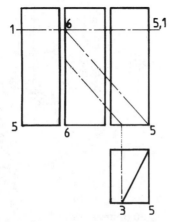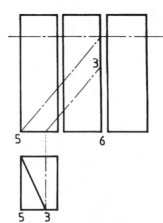

**Figure 3.55**

**Figure 3.56**

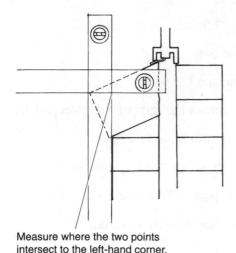

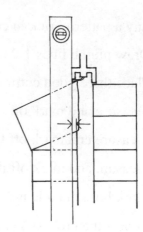

Measure where the two points
intersect to the left-hand corner.

Return corner sills are not difficult but where there will be aluminum windows on each side there can be difficulties in holding the bricks in place as you build the sill. I create a temporary guide as shown in Fig. 3.57. I use two pieces of ¾" × ¾" (19 × 19 mm) timber and clamp them to the frame of the window. Cavity ties hold the corner in place while laying, raking, and cleaning. Because I sometimes think this construction requires three hands, the timber allows me to position the bricks in place. When the mortar sets the timber is removed and the aluminum corner moldings finish the job off.

**Figure 3.57**

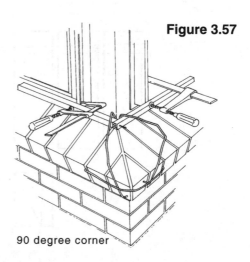

90 degree corner

# 135° Return Corner Sill

1. Bed temporary brick as pattern brick [Fig. 3.58 (3)].

2. Draw in Squint Angle and Bisect Angle in plan [Fig. 3.58 (1)].

3. Draw line 90° to face line [Fig. 3.58 (1)].

4. Transfer measurements from pattern brick to Fig. 3.58 (1) AF and AE.

5. Project these dimensions across to bisecting line to find size of cutoff sections C and D [(Fig. 3.58 (1)].

6. Stand two bricks soldier like and mark off top and bottom slope lengths [Fig. 3.58 (2)].

7. Measure distance C and D from Fig. 3.58 (1) and transfer to Fig. 3.58 (2), producing line from intersection to corner arms showing top cut.

8. Draw plumb cut to bottom slope.

9. On the header face mark in distance D and produce to top arms to complete cut [see Fig. 3.58 (2)].

10. Reproduce opposite cuts on similar bricks.

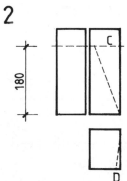

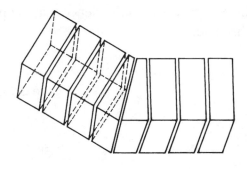

**Figure 3.58**

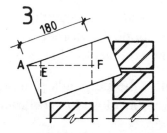

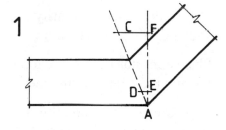

This diagram shows that on 135 degree return corner sills we have to cut out the top edge of the sill brick, otherwise the ribs in the molding will push the corner sill bricks down upsetting the line of the work.

**Figure 3.59**

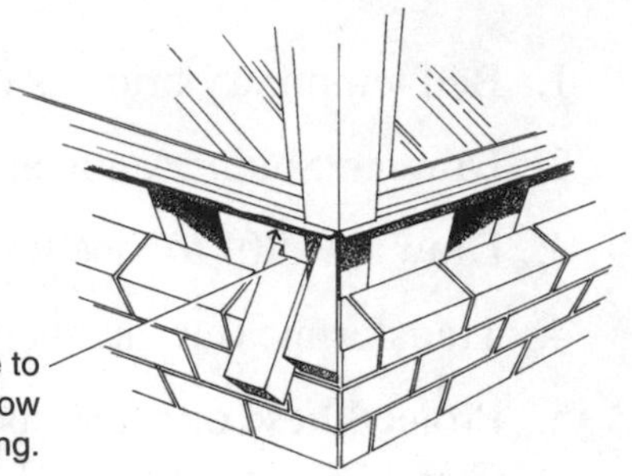

Cut this edge to allow for window molding.

# Toothing Corners

Toothing corners simply means erecting a corner so the bricks of the wall can be interlocked as the wall construction proceeds.

Where possible try to tooth all corners right out at least one and a half bricks. Toothing the corner too small weakens a corner. I have found that anything less than one and a half bricks makes the corner too weak, and this can mean trouble when the line and block is attached and tightened. It could easily pull the corner over.

The same is true for internal corners. Not toothing the full two bricks from the corner makes it harder to fix the line pin in the corner. It's my experience that the pin can dislodge the internal brick if it's just freshly laid.

### Toothing a Corner

1. Place a half bat temporarily on top of the stretcher brick underneath.

2. Place a mortar bed on the top of the half bat and the adjacent stretcher.

3. Lay the header and the stretcher as shown and level those bricks. Now sight down as shown by the arrows. The toothings can be frequently checked by sighting opposite corners as indicated by arrows A and B in Fig. 3.61.

**Figure 3.60**

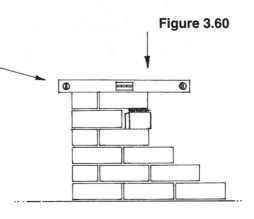

Keeping the perpends plumbed will prevent interference with the fitting of windows or doors close to the corner. (Refer to point C. in Fig. 3.61.)

**Figure 3.61**

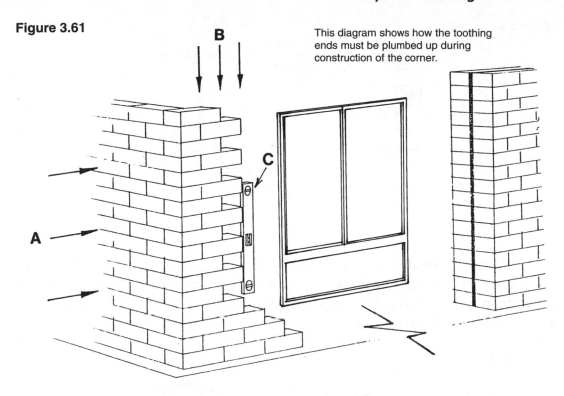

This diagram shows how the toothing ends must be plumbed up during construction of the corner.

Use the same method of sighting for toothing corners as was taught for building up corners. All toothing ends or corner returns must be sighted through to the other corner.

If there is no corner to sight to, sight directly down to existing brickwork. (With practice you'll become an expert.)

In diagram above, "A" indicates points of corner to be sighted through, and "B" indicates points of corner to be sighted down.

**Figure 3.62**

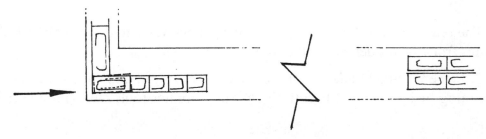

**Figure 3.63**

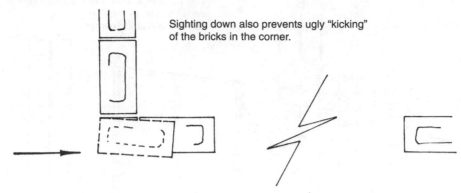

Sighting down also prevents ugly "kicking" of the bricks in the corner.

1. I can't emphasize enough how important it is when laying bricks into the toothing, that they have to be properly grouted with mortar. When laying a brick into the toothing, the brick is held high while being pushed into the toothing, squeezing the perp joint to the correct size.

**Figure 3.64**

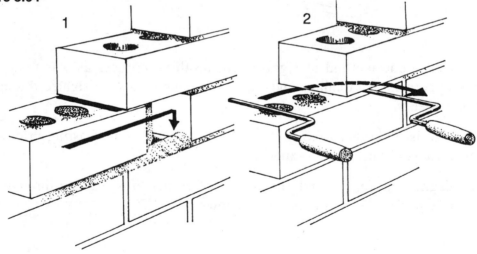

The brick can now be lowered by pressing downwards to the stringline. Fill the toothings from the header end of the stretcher.

2. Place a suitable amount of mortar on your trowel and slide the blade into the toothing bed joint and fill with the toothing jointer tool, removing the trowel

slowly. Make sure the joint is filled to capacity.

I don't like toothing at the best of times, but if they are filled properly there is no comeback.

There are some building codes that prohibit the use of toothing brickwork, mainly because there is concern that the bricks aren't filled properly in the case of frog or extruded bricks.

Toothing is a universal application on brickwork, I feel, and can be successfully done.

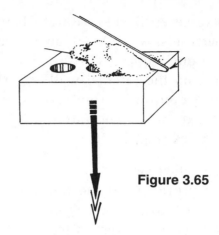

**Figure 3.65**

**Figure 3.66**

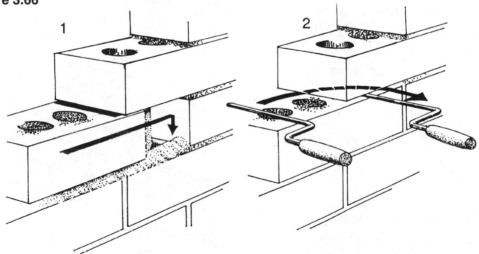

When you are toothing in bricks over courses with extruded holes, the easy solution is to solidly fill the holes of the underlying brick with mortar. In this way, when you backfill the bed joint your mortar won't disappear down the extrusions.

# Bonding Internal Walls

I've never liked to see walls butt jointed either on the run or where they tee off other walls. A crack will always result. We use two methods. On the external wall we cut a queen closure and then on the internal wall we cut a couple of smaller closures to house the intercepting brick (Fig. 3.67).

The other method we use, which I prefer, is to use the diamond saw to cut a check in the brick which will act as the receptacle for the intercepting brick (Fig. 3.67 B).

Whether you cut out the internal closure blocks or the check into the receptacle brick, remember to allow for a ⅜" (10 mm) mortar joint on each side.

In constructing a wall into this wall you will be giving a nice even finished off look. Be sure to plumb all the way up one side so the bricks will fit, while still maintaining plumb on the wall run.

**Figure 3.67**

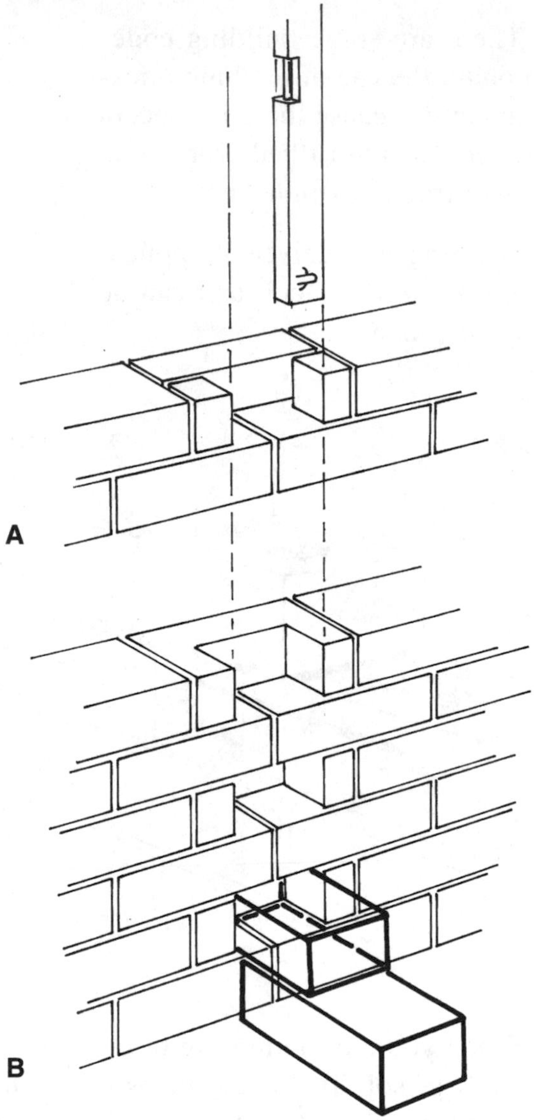

A

B

# Tingle

**Figure 3.68**

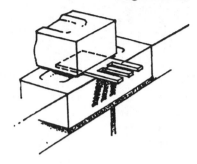

A tingle is cut from a piece of flat metal and is used to hold the stringline plumb in windy weather, and also to correct any sag in the line over long distances.

Generally this tingle should be checked by sighting for level and plumbed and gauged frequently, especially in windy conditions when it should be plumbed with a level.

The tingle person should work in cooperation with the cornerperson, making sure the plumb of the tingle and the cornerperson do not lean in opposite directions creating a bow in the brick mass. A bow can sometimes occur if the stringline is allowed to run out at the ends, and the tingle is out of line with the brick ends. (See Fig. 3.70.)

When working to window heights on a long wall with several windows, the tingleman has the same responsibility as the cornerman in ensuring that the bricks remain in-coursing at the tingle as at the ends. To clarify this you should imagine a 60' (20 m) long wall with a few windows at either end and in the middle. When a couple of layers are working from either end and there is a layer building up the center, it is easy for the middle or the ends to get out of kilter. Even a few sixteenths of an inch will eventually mean problems in meeting window heights consistently with the lines of courses.

Some of our constructions have very long runs and when we have seven to eight bricklayers on the line and with one person required to continually move up the tingle, valuable time can easily be lost. For this work I like to set up a profile as shown in Fig. 3.69. I mark the gauge down and then hook the line on the profile with a door frame

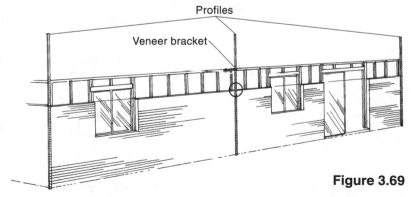

Profiles

Veneer bracket

**Figure 3.69**

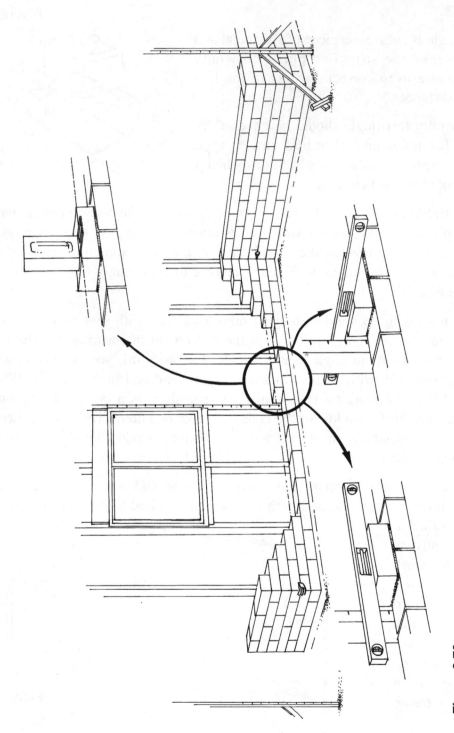

**Figure 3.70**

clamp. This allows the line to be lifted up at every course, level, plumb and to gauge. It's a useful method with long retaining walls of up to 240' (80 m) in length which can be easily laid without constant movement of the tingle.

I would plumb the two end profiles, then using a stringline (only as a guide) I would set up more profiles at 30' (10 m) intervals on very long walls. By sighting through them then setting up a stringline on these profiles with door frame blocks, it allows the stringline to carry right to the end of 240' (80 m) and for there to be a tingle every 30' (10 m). Because 30' is the longest distance I would pull a line without a tingle, shoot a level on all the profiles and mark the bond through before laying.

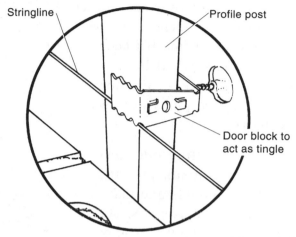

Stringline

Profile post

Door block to act as tingle

**Figure 3.71**

# Piers

When constructing piers, sight across on top of the bricks between the two parallel stringlines and across the face instead of using a level to check for square and level. The construction of piers means correction or checking with a level is difficult because they have four sides and therefore eight plumbing points.

Usually there will be a number of piers in a line. Use the same method as in the building of a corner by sighting down on top of the piers allowing the use of stringlines. Unfortunately when you are working on a single pier you will need to check those eight plumb points with a level as you work up. When building piers, the most common problem is the "growing syndrome" as the perpends have a tendency to get wider. It is a peculiarity of their construction; however, it can be overcome with the following method.

Trying to plumb a free-standing pier on four sides—meaning eight plumbing points—and keeping all correct is near impossible, but can be corrected by a measuring system. I find that it is best to plumb two sides vertically and then measure across two sides at the top corners as you work up.

If the width of the pier varies then runout is evident as the piers are "growing." I find if I keep the widths consistent at the chosen width—say 11¼" (286 mm) for most piers—the joints will always stay the same.

When setting out piers make sure you run the tape over the diagonals and make any necessary adjustments. The diagonals must be the same or the pier will not be square. Also notice how the stringline is parallel with the existing brick wall. There is nothing worse than bricking up to soffits and finding that the pier is out of square or doesn't line up with the existing brickwork.

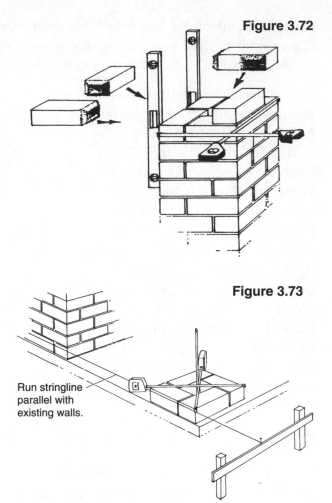

**Figure 3.72**

**Figure 3.73**

Run stringline parallel with existing walls.

**Figure 3.74**

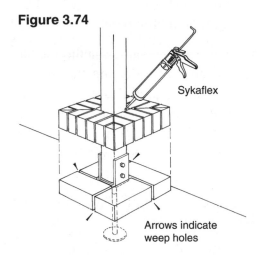

Sykaflex

Arrows indicate weep holes

When using a half-height pier to brick around a steel or timber post, it's important to leave weep holes either at the bottom of the pier or on top of the DPC or termi mesh. Make sure to cut the cap and allow a gap of at least ⁵⁄₁₆" (8 mm) around the post, then hose out around the post, washing all the dags off the inside of the pier. Any moisture that gets into the post will cause the hardwood to expand and cracking will occur. The same with steel on hot days, thermal expansion can again crack the brickwork. After hosing out run a bead of flexible chalking around the cap and the post to prevent moisture penetration.

# Bonding Engaged Piers with Queen and King Closures

There are three main types of piers: freestanding; engaged piers which are locked into the passing wall; and attached piers which use a brick tie wire or cavity tie to attach the pier to the wall. In this segment we are dealing with the engaged piers using queen and king closures. King closures are checked in about one-half of a brick width and length, while queen closures are bricks split along their length and laid horizontally into the wall courses. Queen closures are mostly used when engaging piers into double-brick walls.

**Figure 3.75**

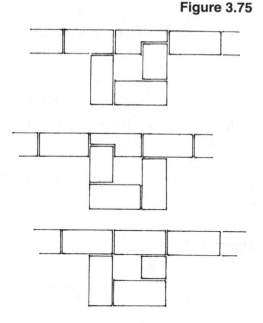

Start your bond on the pier as shown and engage the piers as you construct the wall. Leave the two gaps in the wall alternatively every three courses, to tie both sides of the pier. The diagrams illustrate this a little more clearly.

**Figure 3.76**

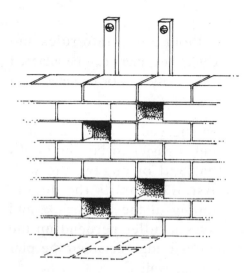

Make sure you plumb the closures vertically with the cross joints so excess filling is avoided when fitting in the "locking" bricks.

Not plumbing the closures correctly leads to trouble in the construction of the pier if the perpends on the wall courses are not lined up, and will mean either very big perpends or no perpends either side of the pier. It could even make it impossible to tie in the pier.

After the closures are laid and engaged into the wall, filling the toothing on top of the tie brick is most important. We discussed this in the segment on toothing of corners.

It is important that you learn to use and enjoy working with the toothing filler tool as it will become a big part of your life as a professional.

Plumb two sides of the pier and then measure two sides as learned before to keep even perpend joints. This method can become very fast, and after practice and a short time, very practical. Engaged or attached piers can be measured the same way.

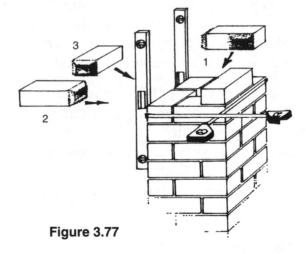

I find it easier when laying freestanding piers to lay the header first, thus giving a full brick perpend to push against the header. The last brick will have to be double-tipped with mortar on the side and of course at the end.

**Figure 3.77**

**Figure 3.78**

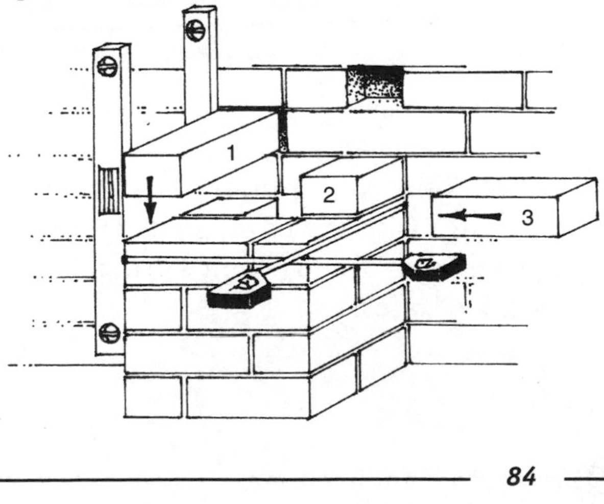

Without going into rules and regulations, piers can be attached with cavity ties or engaged with closures.

On engaged or attached piers I find it easier to lay the bricks that butt or engage the wall first. This omits the double tipping of the last brick, and gives a fuller perpend up the internal edge between the pier and the wall.

These sketches show bonding of a 11¼" (285 mm) engaged pier bonded into a 7⅝" (190 mm) wall, showing two courses of stretcher bond.

**Figure 3.79**

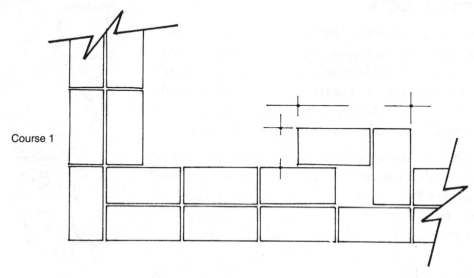

Course 1

**Figure 3.80**

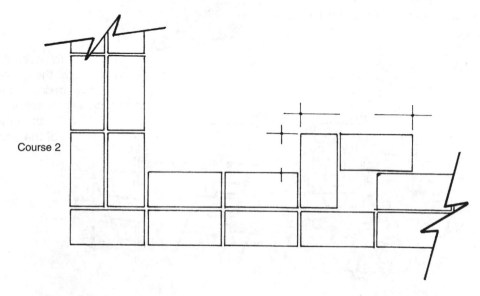

Course 2

# 4

# Scaffolding

## Introduction

Scaffolding is one of the most important aspects of safe and efficient bricklaying especially where the work will exceed 8' (2.4 m) in height. In demanding economic times all service providers must be on the ball. Scaffolding right behind the bricklayers as each wall progresses to scaffold height saves time and money.

Figure 4.1 shows a brick gable end. Notice how the "hop-up" scaffold is set up below the primary scaffold? This allows the laborers to permanently feed the bricklayers with bricks and mortar, keeping progress lively. We have also set up the middle scaffold high enough to reach in one lift.

When loading brick stacks make sure to leave enough room for dippers of water and toolboxes. This saves unnecessary movement of the tradesmen.

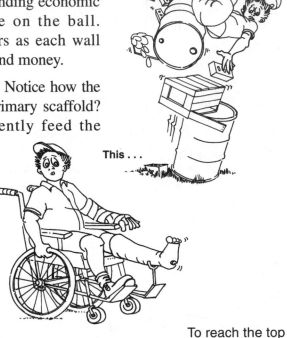

This leads to . . .

This . . .

Figure 4.1

To reach the top of the gable, the middle scaffold is set up higher than the end of the scaffold.

Hop up scaffold.

Figure 4.2 shows the scaffold set up in a way for optimum performance. All scaffold should be four planks wide as this allows enough room for the bricklayers to walk around the back of the stacks of bricks and to operate a second brick cutting saw which can be set up on additional planks. I find the addition of a second saw is almost imperative in my work, as fine-trimming of the bricks can be done by the bricklayers themselves which saves time.

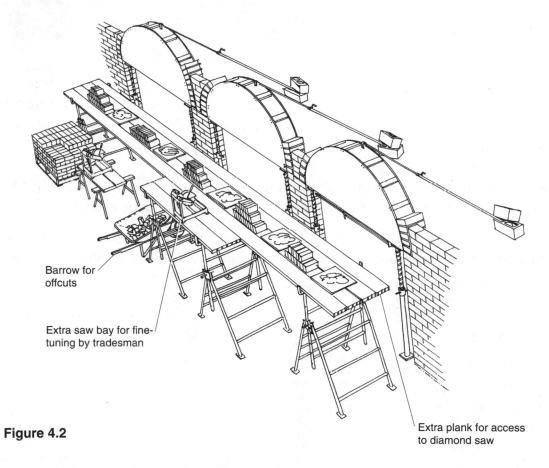

Barrow for offcuts

Extra saw bay for fine-tuning by tradesman

**Figure 4.2**

Extra plank for access to diamond saw

I also place a wheelbarrow in between or next to the saw, into which the offcuts can be thrown.

When setting up scaffolding the importance of safety can never be overstated. Even on lowset, run the scaffold as level as possible and when setting up small trestles up to 5' (1500 mm) high, only pack up under the legs to a height of three bricks. Anything over this is unsafe. A little bit of shoveling can also work wonders to get scaffolding and trestles level. In difficult circumstances we can "pier" under the legs and use planks.

In Fig. 4.3 I have shown one way to achieve some height on a job site. Putting the legs directly on the piers is asking for trouble so I stabilize the setup by putting timber "ties" across the piers, onto which the planks are laid and the trestles erected. When erecting scaffold it is important to give some thought to safety. If it's not a structure that you would feel confident working on then never allow anyone else to work on it either. My suggestion is to help out on those jobs where scaffold may not be available.

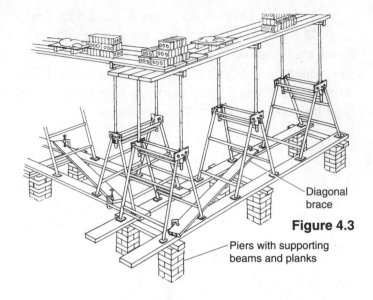

Diagonal brace

**Figure 4.3**

Piers with supporting beams and planks

For highset buildings built two stories, I erect "outriggers" to the scaffolding. The internal rigs, on which the layers work, are two planks wide and the external outriggers on to which the bricks are loaded are three planks wide.

I must emphasize that setting up the scaffold beforehand and loading the scaffold with bricks and equipment could be the making or the breaking of a successful commercial job. When the jobs are lean on profit the only solution is to get the job out of the way as quickly as possible.

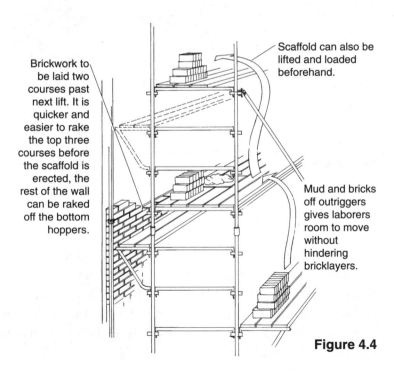

Brickwork to be laid two courses past next lift. It is quicker and easier to rake the top three courses before the scaffold is erected, the rest of the wall can be raked off the bottom hoppers.

Scaffold can also be lifted and loaded beforehand.

Mud and bricks off outriggers gives laborers room to move without hindering bricklayers.

**Figure 4.4**

The next lift can be set up in advance with bricks and mortar when the wall is laid to scaffold height. Only the inside "outriggers" have to be shifted as shown by the dotted line in Fig. 4.4. This method can be used on a lowset brick house. Notice how the inside planks are lower, and the planks that the mortar and bricks are loaded on are at a more accessible height, creating a quicker and faster working environment (Fig. 4.5).

Outside planks set up higher for speed and performance

Two inside walking planks set up at comfortable finishing height

**Figure 4.5**

# Using the Brick Elevator

First meet the brick supplier on site when the bricks are delivered and stack the pallets together to allow all the bricks to be blended as required. The bottom of the elevator is kept mostly stationary. The elevator's position is universal and can be positioned directly over the pallet or anywhere around the pallet.

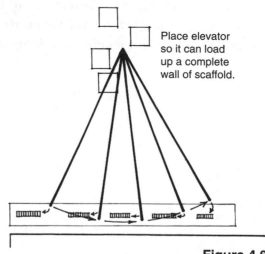

Place elevator so it can load up a complete wall of scaffold.

**Figure 4.6**

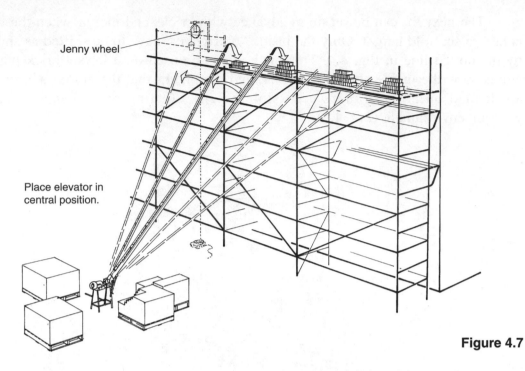

Jenny wheel

Place elevator in central position.

**Figure 4.7**

Using the elevator I prefer to finish one complete side in one day as all the cleaning down only has to be done once. When stacking bricks with the elevator give some thought to loading off the elevator. I work across the plank from one end sliding the elevator along as I progress. This is much easier than having to lift the elevator over a stack of bricks on the plank.

*Always be alert to the hazards of working with elevators. Long-sleeved shirts can become entangled, as can the cords on hats. Elevators, like all mechanical and electrical equipment, require care in operation.*

# 5
# Cavity Brickwork

## Introduction

There are two types of brick construction. Cavity brick means two walls of brick running up side by side. This double skin of bricks is used for strength, but it also has insulating qualities because—much like a thermos or esky—there is a cavity between the walls which will trap and maintain air temperature in both summer and winter.

Cavity brick houses are more expensive because there are twice as many bricks in the job, and not a great requirement for timber framing and internal sheeting. In the long term however, they are an investment, because the house is constructed completely from a durable and serviceable building material that never rots, is not affected by weather, and never needs painting. They are cooler in hot climates—insulating from the outside penetration of the sun—and warmer in cool climates because they maintain temperature.

Brick veneer houses are single-skin constructions that rely on a secondary building product for the internal walls and cladding. These houses are extremely serviceable and popular.

When setting out for the cavity wall lay the inside walls first, fitting windows and damp courses as required and any plumbing or wiring. Be sure to close the cavity between the walls by moving the inside skin outwards ¼" (5 mm). This makes the cavity slightly smaller—from 9" (230 mm) to 8¹³⁄₁₆" (225 mm). This is mostly done to overcome difficulties with the construction style of aluminum frame windows. Doing this, the windows can be set up 3¼" (83 mm) off the face of the inside skin, thus giving the window cover on the outside, without returning the brick or cavity closures.

Try not to make the cavity too small, otherwise bonded returns where the outside skin returns to meet the inside skin will cause problems. It will mean that the cut bricks will probably have their cross joints almost perpendicular where the outside skin returns and is bonded to the inside skin.

Too many vertical joints in line and close together are unsightly and structurally unsatisfactory.

Beginning as mentioned with the inside skin, place a cut on all corners so the bond will correspond with stretcher bond on the outside brickwork.

This eliminates cutting on either skin around windows.

**Figure 5.1**

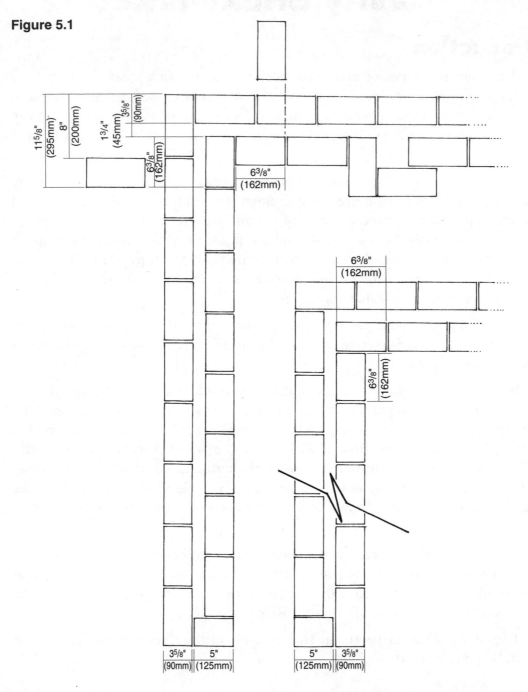

Remember also that the bond on the inside must be opposite the bond on the outside.

If the cavity is 9" (228 mm) put 6⅜" (160 mm) cuts in the corner. The bond on 9" (228 mm) cavity work will be opposite bond on the two skins.

When scraping the back of the inside wall, place strips of masonite over the cavity between the skins to keep extraneous material from falling and filling the cavity. It could cause water problems.

**Figure 5.2**

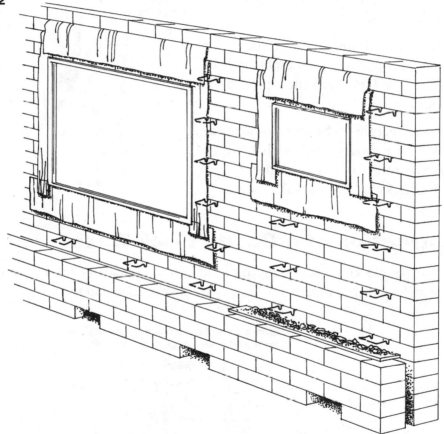

On the outside skin, leave out an occasional brick along the bottom course so the wall can be hosed out each day after work to eradicate the excess mortar.

The courses of the outside skin should never rise above the inside skin because water can then flow down the ties into the inside wall and present permanent water problems for the owner.

To waterproof steel door frames, run a bead of silicone as in (Fig. 5.3A) vertically down the rebate in the frame then cut a piece of 6" (150 mm) DPC approximately 10" (250 mm) longer than the frame and fit as shown in (Fig. 5.3B) I find it much better to fit the DPC to the frame the day before using a piece of timber and some G-clamps. This gives it a chance to stick. If the ties (Fig. 5.3C) are small they can be slotted through the DPC.

This method allows the frame to be filled and the door tie to be embedded in mortar and the DPC will then provide a waterproof barrier.

Where there is to be quite an amount of fill behind a cavity brick wall under a poured concrete floor it is important to ensure the brickwork has lateral strength.

**Figure 5.3**

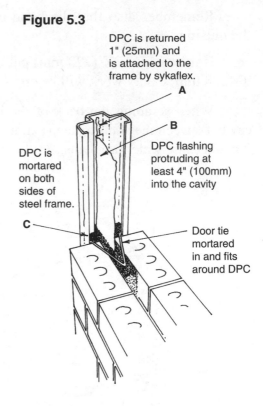

DPC is returned 1" (25mm) and is attached to the frame by sykaflex.

A

B

DPC flashing protruding at least 4" (100mm) into the cavity

DPC is mortared on both sides of steel frame.

C

Door tie mortared in and fits around DPC

# Construction of a Cavity Brick Base

Most people know that many more than four courses of bricks won't stand a lot of "push" against it. Figure 5.5 shows the cross section and the completed job.

**Figure 5.4**

Cavity base — it is much easier to lay the inside skin first from the outside where possible.

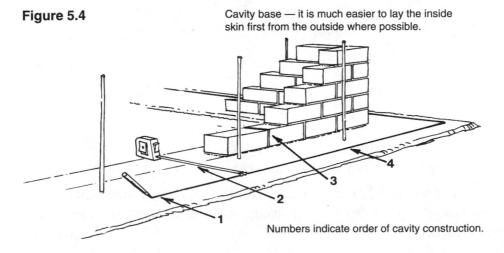

Numbers indicate order of cavity construction.

Figure 5.4 shows how we set up the job. Run all the string lines from the profiles around the base so you can plumb down to the footing. Mark it out with a waterproof pen so the marks will withstand foot and barrow traffic. It's easier to build a wall from the inside. Now measure back from the line 5⅞" (150 mm). This is where the inside courses will run and allows one skin of bricks and 2¾" (70 mm) for the cavity that will be filled with concrete.

Build up corners and run the brickwork between allowing cavity ties every three courses. Now move on and build up the outside skin starting with the corners as shown in number 4 and run up the outside brickwork.

In Fig. 5.5, notice how the outside skin is run higher to allow for a thickening beam. Figure 5.6 shows typical cavity brickwork with beam, compacted fill, waterproof membrane, steel, concrete cavity fill and floor all in one—and that provides excellent strength.

Shows steel and floor beam.

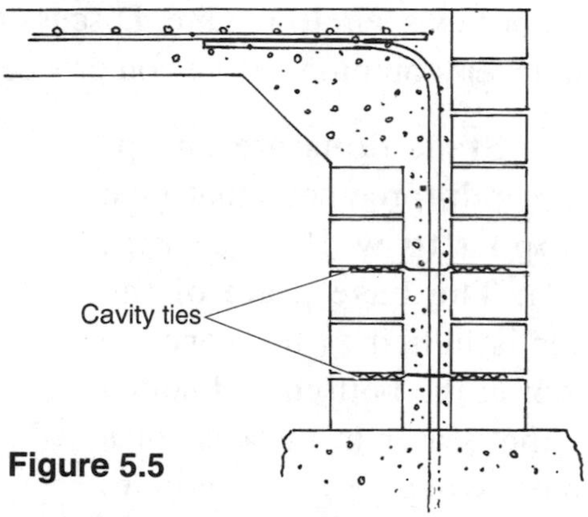

Cavity ties

**Figure 5.5**

**Figure 5.6**

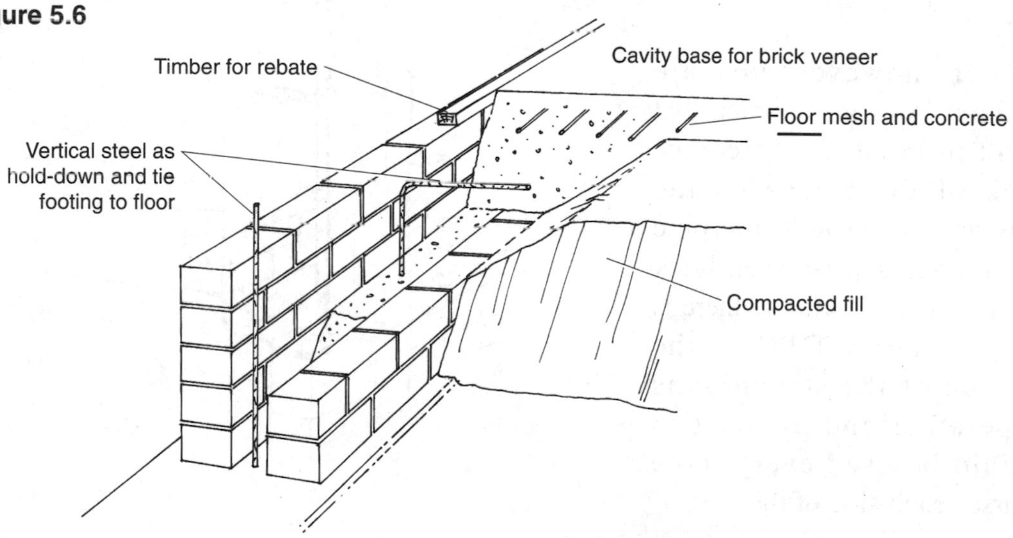

Timber for rebate

Cavity base for brick veneer

Floor mesh and concrete

Vertical steel as hold-down and tie footing to floor

Compacted fill

# Laying Cavity Brickwork Structurally

As a bricklayer I've always been a very strong advocate for brick construction. I like timber and the various other materials, but purely as a practical value for money building material, I prefer bricks. However everyone acknowledges there are some considerations in getting brickwork to meet all local authority requirements, especially in coastal and cyclonic areas.

The following ideas are my suggestions to help people who want a complete brick home, but can't for one reason or another satisfy local building requirements. These ideas are intended to overcome problems with wind resistance and structural strength required by bracing walls and the various other nuisances that building authorities seem to require. Discuss them with your architect and builder and get an engineer's opinion before you proceed!

Stanchions are an upright reinforcement that give support to walls (see Fig. 5.7). The base plate of the pier is bolted to the concrete floor at the bottom and bolted to the bearer in the case of a two story brick house. Galvanized steel is preferable and should fit between the brickwork.

If, however, you are required to fit a stanchion that won't fit in the cavity, cut the back of the bricks on the outside skin around the pipe and fill the gap between brick and pier with mortar so there is no movement. DPC on the outside of the stanchion is imperative and cavity ties should be used every three courses each side of the post.

**Figure 5.7**

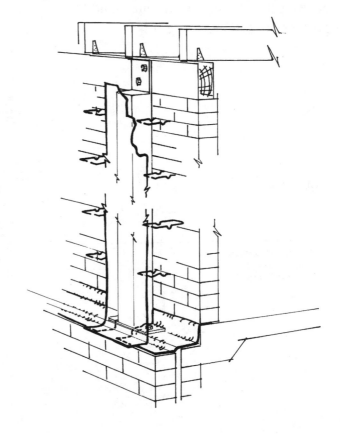

On a lowset cavity brick dwelling opt for a bigger top plate — as big as you can get away with. Tie the top plate into the foundations as per diagram at right.

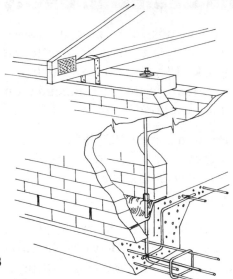

**Figure 5.8**

In Fig. 5.9, I offer some suggestions for bracing external brick walls. Run a row of pavers vertically between the skins of brick. Most pavers are 1⅜" (35 mm) thick, making them ideal to fit in a 1½" (38 mm) cavity. Wherever internal walls are timber, use bracing ply for structural strength. Be sure when you stack the pavers that you get a good mortar joint between them.

Be sure to tie it all together by using cavity brick ties, each three courses on either side of the pier. And don't forget the damp course. (Good luck with the authorities, and remember the old saying, "There's never any reason for it — it's just their policy!")

**Figure 5.9**

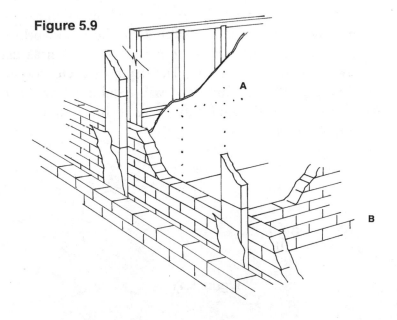

# Bond Beams on Cavity Brickwork

This is possible—indeed, easy—to do and adds structural strength to the brickwork. If you want an all brick dwelling, here is my recommendation for including a bond beam.

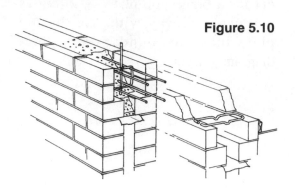

**Figure 5.10**

Position some DPC [4¼" (110 mm)] across the cavity of the brick skins under the last course before you build the bond beam.

Then lay two courses of queen closure bricks on top of the last course of bricks over the DPC. Between the two courses of closures place long cavity ties to tie the courses together. This will prevent them spreading when the bond beam is filled with concrete. Fit all necessary steel and reinforcing as required (refer local authority details) and backfill with concrete. I personally feel that combining this method of bond beam construction with a steel stanchion should get you permission to build your house in any wind-rating category.

In the figure below I've detailed similar all-brick construction which includes the use of brick piers which are sealed by the vertical pier paver method and backfilled with concrete. The piers can be internal or external.

Cavity brickwork is not regarded as having lateral strength. Therefore, by using cavity piers, we can strengthen the brickwork without bulky and space-consuming "engaged piers." Lay the inside skin as usual placing piers besides windows and doors. The depth of each pier should be a minimum of 9" (230 mm) and will require cavity ties on both sides. The ties not only serve to brace and strengthen the pier but also hold the brick closures, which are laid in the cavity to retain the concrete fill.

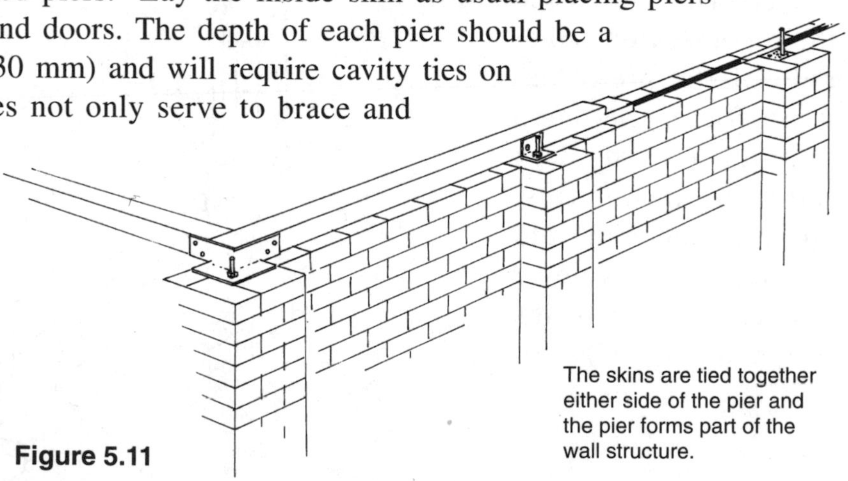

The skins are tied together either side of the pier and the pier forms part of the wall structure.

**Figure 5.11**

Another method is to nail timber vertically up the wall on either side of the pier. Be sure to make the width of the timber smaller than the cavity, as bricks can vary in width up to ⅜ (10 mm) and you could end up cutting the backs of the bricks around the timber. It is easy to run a mortar joint up the timber, thus sealing off the cavity.

I like to place the ties every three courses as this prevents the brick skins from "blowing out" during filling with concrete. Notice the DPC protruding either side of the closures and the cutting around the ties and the sealing with silicone to prevent any moisture penetration.

Next insert the vertical steel depending on the wind loading for the location. In a W40 category area, which is general, low wind loading, I would use two ½" rods and one ½" threaded rod as a tie-down for each pier. It's also easy to add more steel to accommodate greater wind loads. For example we could duplicate the requirements for 6" (150 mm) blocks, (W50 wind category) two ⅝" rods. Use a threaded ⅝" (16 mm) rod beside each opening wider than 6' (1800 mm) and at 3' (900 mm) centers between.

For extra strength you could tie the vertical cavity piers into a brick bond beam as in Fig. 5.12 below.

Constructing the bond beam, I prefer to use matching thin wall bricks which saves a lot of cutting closures that are hard to hold up and don't remain in place as well when the concrete fill is poured. This beam also allows L bolts to be fitted as hold-down bolts.

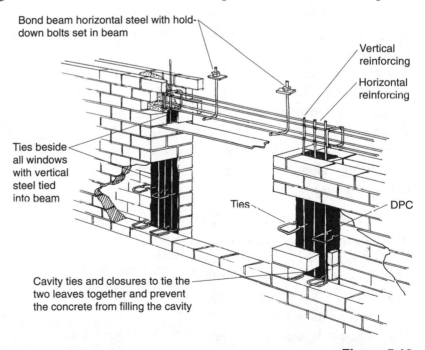

Bond beam horizontal steel with hold-down bolts set in beam

Vertical reinforcing

Horizontal reinforcing

Ties beside all windows with vertical steel tied into beam

Ties

DPC

Cavity ties and closures to tie the two leaves together and prevent the concrete from filling the cavity

**Figure 5.12**

# Cavity Brickwork – Used to Retain

A 3½" (90 mm) cavity is the minimum you should use for any brickwork that has a retaining job to do. There will be engineering standards that must be worked to, but with a bit of planning you can construct a mighty strong retaining wall that will have excellent retention properties. I am confident a 3½" (90 mm) cavity wall, when properly constructed, will retain to very pleasing heights [up to 8' (2.4 m) in fact] without the expense of concrete blocks, wider footings, or extra concrete. Because of the 3½" (90 mm) cavity you will need extra-long cavity ties each three courses. These ties are also useful to lay the horizontal steel on during construction.

"Cleaning eyes" left each five or six bricks, which can be fitted later after the hosing out of the cavity, will ensure you get a good fill with concrete. In considering the holding capacity of cavity or any masonry construction, most are designed so that when the concrete floor is poured, everything will be tied together and this is where most of the trouble can begin. When compacting the fill you must back-prop the walls. Talk to your engineer but having a lot of experience in these matters, I find that lots of "bowing" and "bending" of the masonry will result and this makes "propping" during backfilling, imperative especially on walls of 6' (1.8 m) and above.

Most of the trouble begins at 20" (500 mm) or over because at the bottom of the wall all we have is compression. To build up the strength internally we can engage piers as is seen in Fig. 5.13 and place a bond beam around at a suitable height to stop the brickwork from bowing out. Placing ⅝" rods at 8" (200 mm) centers gives the cavity enormous strength and saves back propping.

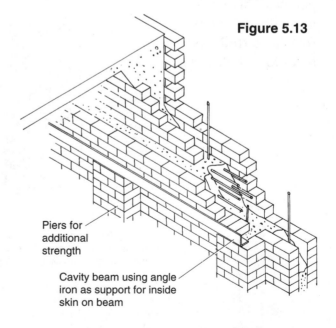

**Figure 5.13**

Piers for additional strength

Cavity beam using angle iron as support for inside skin on beam

When constructing the bond beam we can corbel the brickwork out in a few courses (see Fig. 5.14) or use angle steel 4" × 4" × ³⁄₁₆" (100 × 100 × 4 mm) as shown in Fig. 5.13. Place the angle across the piers or prop it with timber from underneath. The steel provides only temporary support until the concrete in the beam cures.

**Figure 5.14**

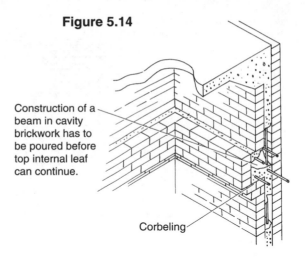

Construction of a beam in cavity brickwork has to be poured before top internal leaf can continue.

Corbeling

## Using Blocks as Beams Inside

As blocks are 7⅜" (190 mm) wide with a cavity of 3" (75 mm) we can use them effectively for our beam. The brickwork is run as before but we lay a course of E blocks on top of the existing cavity brickwork. We use E blocks because there is no web on the inside and this allows the concrete to be easily poured into the cavity. We then place an appropriate gauge of steel reinforcing, lay a course of knock-out blocks over the steel, and the brickwork can continue. If you choose to fill with concrete at this point don't forget to include the vertical steel. (It's difficult to get it in when the concrete has cured.) Be aware though mixing and matching concrete blocks with clayware can and does cause shrinkage or expansion cracks.

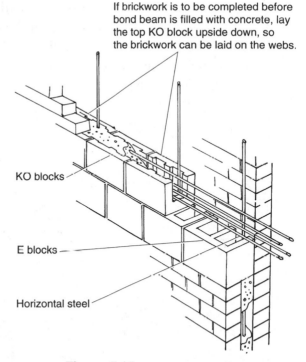

If brickwork is to be completed before bond beam is filled with concrete, lay the top KO block upside down, so the brickwork can be laid on the webs.

KO blocks

E blocks

Horizontal steel

**Figure 5.15**

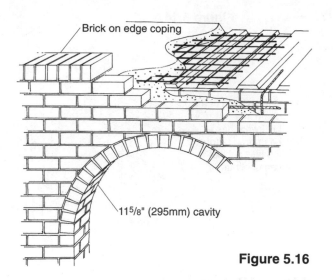

Brick on edge coping

11⁵/₈" (295mm) cavity

In Fig. 5.16 we see the same construction using cavity brickwork with lots of long ties. Refer to Fig. 14.8 for construction details.

**Figure 5.16**

A brick base can be used to retain fill to support a concrete slab. Internally there is blockwork on the bottom side of the building to retain the fill, for economics on the front and low end we use cavity brick. On the front patio which has a timber floor we build up piers to carry a bearer that supports the joists. This area will require vents. Don't forget the access door.

**Figure 5.17**

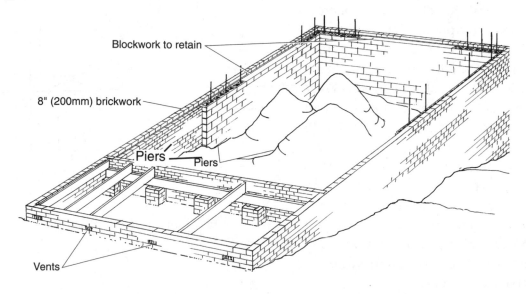

Blockwork to retain

8" (200mm) brickwork

Piers

Piers

Vents

### Brick Vents

Leave a space of one and a half bricks by two courses high at the appropriate spacings (this depends on the area you have to ventilate). These vents are generally placed in the brickwork to ventilate the area underneath timber floor houses, keeping the timber structure dry by

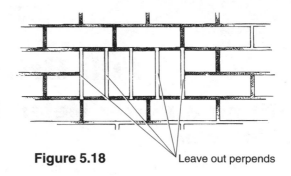

**Figure 5.18**                    Leave out perpends

allowing fresh air in and taking moist air out. These vents don't require painting as do cast or terra-cotta vents. Laid out of brick, they blend in and match the existing color.

# Replacing Cleaning Eyes with Closures

Whenever there is a need to clean out a cavity in brickwork it is of course helpful to have cleaning eyes. When the bricks are replaced however, it is not unlikely that the mortar on the bed joint will be pushed in and could close up the cavity, preventing the weep holes from working. I recommend cutting 1⅛" (30 mm) off the back of the brick closures that are to be fitted. As long as the mortar is packed into the bed joint there should be no structural problem. Again make sure to clean out the weep holes.

**Figure 5.19**

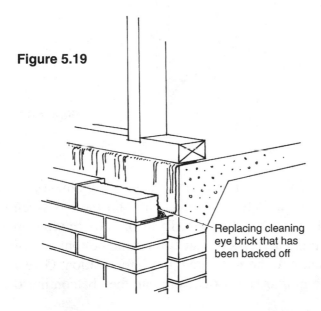

Replacing cleaning eye brick that has been backed off

# 6

# Full Brickwork

## Fitting a Power Meter Box into 8" (200 mm) Brick Wall

Step one in fitting anything into brickwork is to measure it so you know how big a hole to make. Never assume any two power boxes will be the same size.

Electricians prefer the meter boxes be installed in the front third of the dwelling if you are working on a new house. In a double-skin or cavity brick wall, measure the height at which the box will be positioned. In Queensland we fit our boxes between 4'–7' (1.2 and 2.1 m) off natural ground height and try always to work our box into a course of brickwork.

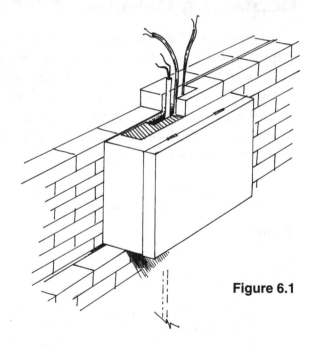

**Figure 6.1**

The inside skin of bricks is completed as normal. Using a level, draw a line on the inside skin up each side of where the meter box will be positioned. Then proceed to brick up the outside skin ending the bricks at the outside line of where the box will fit. I allow about ³⁄₁₆" (5 mm) each side so the box won't be too tight a fit.

Fitting the box after the courses of bricks are laid allows you the advantage of being able to stringline straight across the opening. This prevents errors in course alignment that can occur if you fit the box and try to brick up to it on either side. Fit the box when the outside brickwork is level with the height of the box. To fit the box, mortar up the ends of the brickwork in the opening and lay a mortar bed on the bottom course where the box will sit. Position the box into the wall and tap it into position. Give a few sharp taps down to seat it into the mortar bed. Brickwork can then be continued over the meter box but be sure to allow a chimney or cut-out for wiring and to place a prop in the box so it isn't collapsed by the weight of subsequent layers of bricks.

You should also run a piece of damp proof coursing over the first course of bricks over the meter box between the two skins of brick. Allow two weep holes in the brickwork on top of the damp course.

When fitting a meter box into single skin brickwork try to use a nearby pier to hide the cabling or conduits.

This gives the job a much more professional look. But if you are fitting the box into a pier try to use only a corner of the pier.

**Figure 6.2**

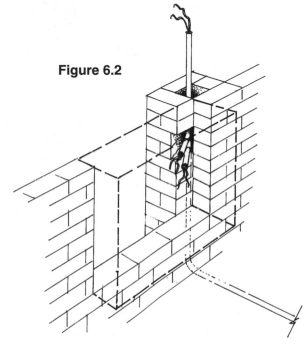

Because the piers are tied to the brickwork with cavity ties you should allow at least the length of one brick to tie into the wall for strength. Be sure to leave a suitably sized hole in the pier for the cables to exit the pier and enter the box.

Bricking up a meter box in a single skin wall is the same as a double-skin wall but you will need to plumb each course vertically with a level because you don't have the benefit of the second skin to mark your vertical lines on.

# Bricking Up to Installed Boxes

When electricians install meter boxes, they never take the time to find out if the height they are fixing them at will work brickwork.

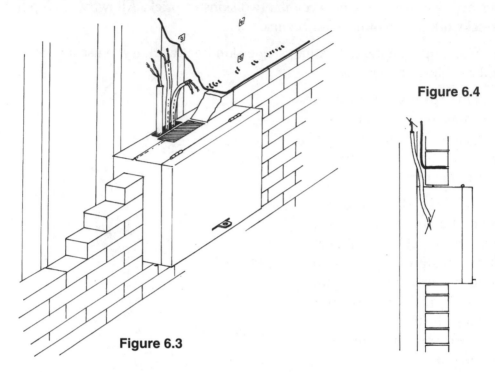

**Figure 6.4**

**Figure 6.3**

While I make this statement tongue in cheek, every bricklayer realizes the difficulties in running brick courses successfully up to preinstalled windows, doors, meter boxes, etc.

Of course the easy way out is to cut bricks "on the run" as they are coursed up to the offending installation. A more professional method, however, is to make allowance for these occasions and to plan your brickwork accordingly. In the case of meter boxes or gas meter installations, I prefer to run my brickwork up to as close as possible under the box. Then I run a vertical plumb line up each side of the box and mark the appropriate gauge all the way up. Next I work away from each side of the meter box—that is from the box out. Having the gauge marked on the vertical overcomes the problem of not being able to run a stringline through.

My preference is also to cut bricks to suit over and beside the meter box. It looks better than soldier courses. It means a little extra cutting work but a job worth doing is worth doing right.

# Using a Meter Box Bracket

I have designed this bracket to fit most meter boxes, electric or gas. A meter box already in situ and cabled is difficult to shift. A lot of time can be wasted building up corners to work the wall up to. Because the box always protrudes past the wall line it's difficult to run stringlines.

When first starting the wall, plumb off the base and draw a plumb line up both sides of the box. Measure the width and the height of the box and adjust the bracket to these measurements and slide it over the box (Fig. 6.5). Place the line holder on

the plumb line and plumb the bracket. Tighten the holding bolts. When the brickwork is up to the underside of the box the line can be fitted directly to the bracket and the gauge can be marked up and brickwork continued.

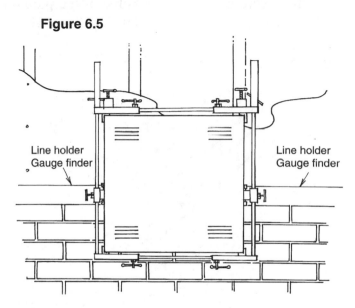

**Figure 6.5**

Line holder
Gauge finder

Line holder
Gauge finder

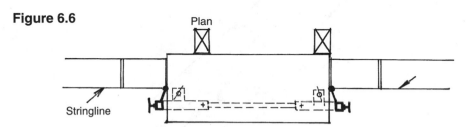

**Figure 6.6**

Plan

Stringline

If there is a big run of brickwork around the construction, have a tie-off point on the meter box side for the line. As the bracket is fitted to both sides of the box this overcomes the need to build corners first and results in a perfectly plumb line of work beside and over the box. The meter box bracket is a handy tool for any contractor and is certainly a good time-saver.

# Fitting Plumbing and Electrical into Full Brick Construction

Cavity brickwork on a dwelling should house as much as possible of the plumbing and the electrical.

Many times I've looked at cavity brick houses where the plumbing has been screwed or fixed to the outside of the brickwork. It indicates a lack of planning and an unprofessional attitude.

To ensure as much as possible of the plumbing and electrical is concealed in the brickwork, a meeting of tradespeople involved in the project is required.

**Figure 6.7**

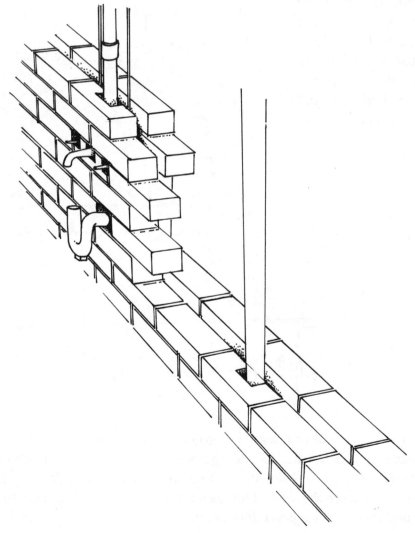

**Figure 6.8**

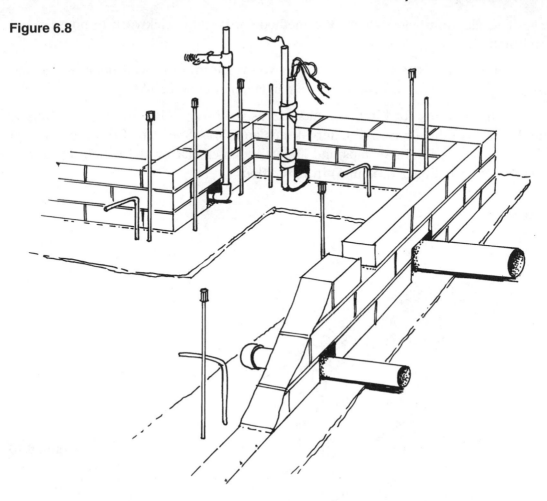

Look over the plans and then plan your work to accommodate the other trades. I like to liaise with the builder on the job to find out about electrical, phone, and plumbing services that will be connected to the dwelling or construction so that provision can be made at ground level for their inclusion on all brick construction. See Fig. 6.8.

Patience and skill with a diamond-toothed saw will allow you to neatly and professionally fit brickwork around plumbing and electrical. Be sure to fill around all plumbing with mortar to stop movement that can cause leaks. In cavity brick constructions I like to build up the inside skin, let the other trades run their connections up and test their services, and then I come back to finish the outer skin.

I also like to conceal all plumbing and electrical when I'm working on single skin brickwork. It's not as difficult as it sounds.

The diagram below shows the method for cutting brickwork to house a sink combination.

It's a bit of extra work but it gives you pride in a job well done. Remember also that plumbing fittings are usually set in at a particular height.

Combinations for kitchen sinks are usually 41¼" (1050 mm), the same as laundry tub combinations. All external taps are 33⅞" (860 mm) from ground level. Sinks and basins that have taps and spouts mounted flat on the bench top usually have the combination 2' (600 mm) from the floor.

**Figure 6.9**

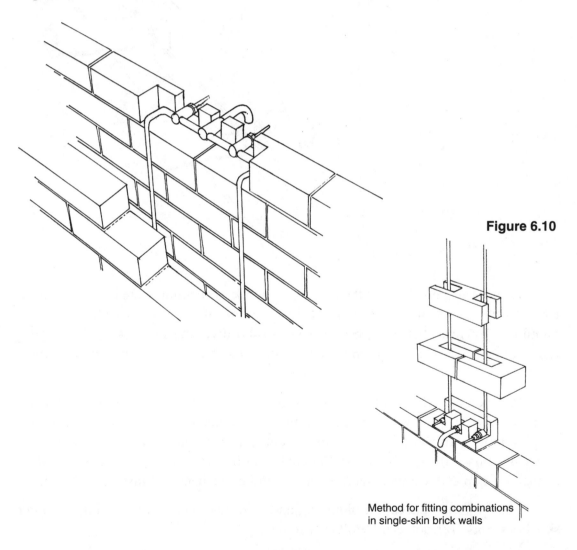

**Figure 6.10**

Method for fitting combinations in single-skin brick walls

In areas where cupboards or tubs hide the brickwork, it is easier to cut or chase the plumbing pipes and electrical into the brick face. If you do let pipes into the brickwork, make sure they are let in past flush with the face of the brickwork so they can be mortared in and tiled over. A bit of planning beforehand can save a lot of time.

**Figure 6.11**

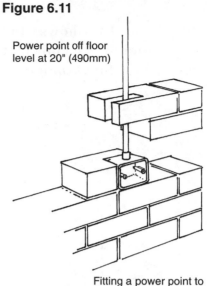

Power point off floor level at 20" (490mm)

Fitting a power point to a single-skin brick wall

**Figure 6.12**

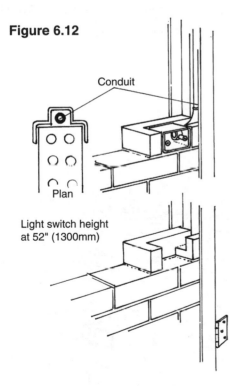

Conduit

Plan

Light switch height at 52" (1300mm)

I've seen fellows cut brickwork all around an electrical conduit when the conduit could have been run down a door frame less than 12" (300 mm) away.

Figure 6.13 shows the method of running conduit that will result in a block-mounted power outlet.

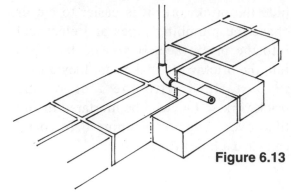

Figure 6.13

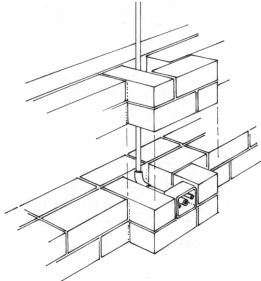

Figure 6.14 shows the same job fitting the block mount into the brickwork so the power outlet has a nice flush finish with the brickwork.

Figure 6.14

Figure 6.15 shows fitting of electrical conduits into cavity or 8" (190 mm) brickwork.

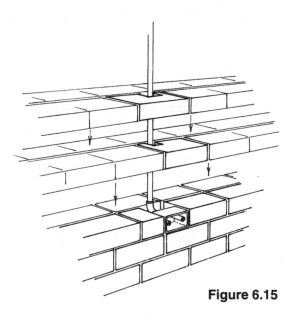

Figure 6.15

# 7

# Miscellaneous Applications

## Curved Walls

At this stage it is relevant to step through the procedure for curved walls, garden beds, fences, and bars, because these will occupy a fair percentage of your time as a brickie.

When constructing curved walls, the first step is to draw the radius on the footing or the floor. The sharpness of the curve will be the deciding factor as to whether the bricks are to be stretcher bond, or built out of half bats, or each brick cut to a taper.

**Figure 7.1**

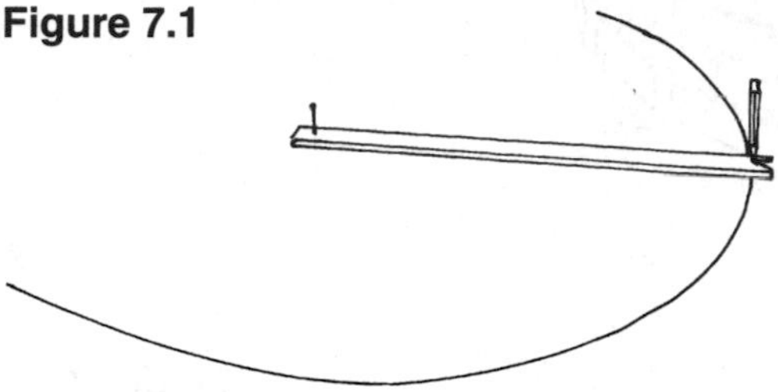

I fasten one end of a piece of timber with a nail and run a pencil around the timber end to mark the radius.

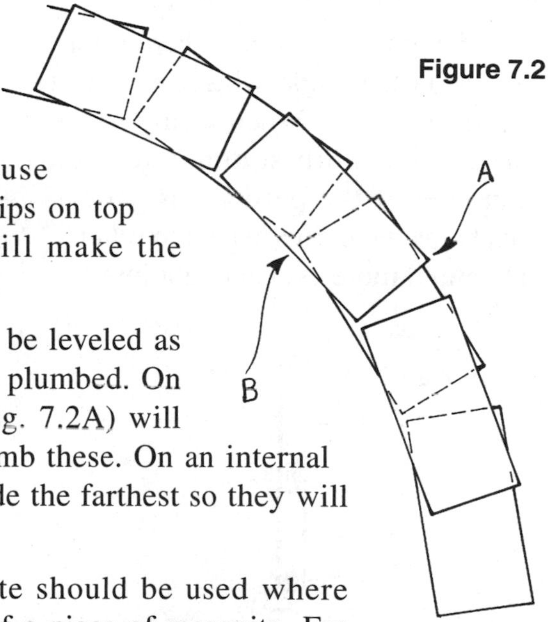

**Figure 7.2**

Lay the first course around the radius. If the radius is too sharp, stretcher bonds cannot be used because lipping of the bricks will occur. These lips on top of each other throw shadows that will make the brickwork appear to be jutting in and out.

Each course of bricks will have to be leveled as it is laid, and each brick will have to be plumbed. On an external curve, because the ends (Fig. 7.2A) will stick out the farthest, it is easiest to plumb these. On an internal curve the middle (Fig. 7.2B) will protrude the farthest so they will have to be plumbed.

For speed and neatness a template should be used where possible with the correct radius cut out of a piece of masonite. For this method there will be three plumbing points; one at the beginning of the curve; one at the end; and one in the middle.

With the course leveled, three points are then plumbed. The template is placed on top of the course of bricks and the bricks adjusted to suit. The use of a radius rod is an alternative method.

**Figure 7.3**

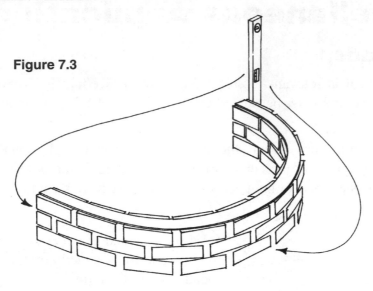

# Squint Corners

Squint corners are where bricks are cut on the angle where they meet to form a corner. When laying squint corners, cut both squint bricks on each side of the perpend as shown. This helps the cornerperson to retain a tighter and more even straight joint.

**Figure 7.4**

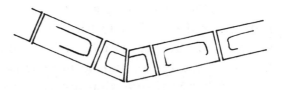

**Figure 7.5**

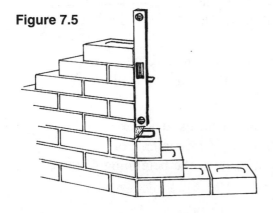

Build one corner of the squint first. This enables easy plumbing of the vertical perpend, and allows the cornerperson to butt the adjoining corner, without awkward and excessive plumbing.

Tie across the corner with a cavity tie or brick wire and keep the perpend joint grouted fully to give maximum strength.

# Laying Hard Up underneath Ant Cap from the Outside

This requires the laying of two courses of brickwork at the one time. By laying the brick which fills the gap to the ant cap on top of the lower course, access is gained to mortar bed and bricks on the top course as the lower course progresses.

Lay the first lower brick to the line by normal method. The next brick will be slid in on top of this brick under the bearer.

**Figure 7.6**

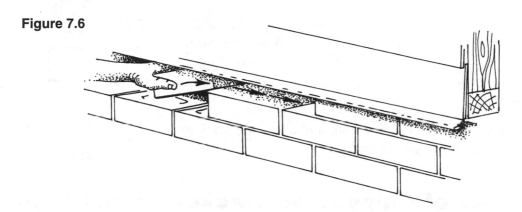

Use the method shown here. This will allow you to place mortar on top of the brick wall in the confined space without dirtying the face brick and wasting mortar. Having placed mortar in position, furrow, then butter a brick, and using both trowel and hand, slide the brick in under the bearer, making sure not to finish with a large perpend when the mortar builds up due to pushing the brick on the bed joint.

When laying the last course in difficult positions, such as hard up underneath a bearer or under ant cap from the inside, sight the underside of the finished material to be laid to. Excess bowing of this structure or frame, particularly down, can be allowed for or corrected from the beginning, preventing excess cutting and saving considerable time.

Lay enough mortar for one brick at a time, on top of the wall, underneath a bearer or ant cap, as the case may be. Using your trowel, flatten the mortar into a wedge shape, approximately ⅜"(10 mm) high with no holes in the bed, ensuring the wedge slopes up in the same direction in which the brick will be pushed.

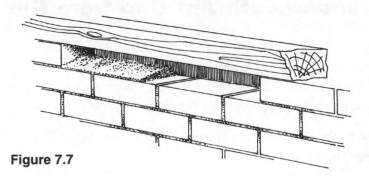

**Figure 7.7**

Butter the underside of the brick as you would a sill brick, in effect with a reverse wedge. While the brick is upside down butter the perpend, and, carefully tipping the brick up the correct way, slide it in on top of the prepared bed.

This will form a full bed joint and allow the brick to fit firmly under the bearer or ant cap.

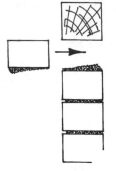

**Figure 7.8**

# Removal of Chipped or Cracked Bricks from Existing Brickwork

There are often occasions in the life of a professional when time will be spent removing bricks from a wall or construction as opposed to putting them in. The procedure I will now describe is for just such times when, for whatever reason, it becomes necessary to remove a brick from a completed wall.

Cutting out the mortar around the perimeter of the brick to be removed creates a cavity which prevents the chipping or breaking of brickwork around that brick.

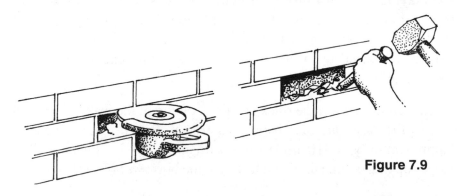

**Figure 7.9**

If a brick is chipped or cracked, first remove all mortar around the brick by cutting the mortar with a power saw fitted with a masonry disc. If this is not possible, use a plugging chisel. Chisel out the face of the brick to a depth of 1½"(40 mm). This will allow you to replace the damaged section of the brick that is visible with a queen closure.

In single-skin brickwork it also helps ensure that the face of the back of the wall will not be damaged. In cavity brick construction, this method prevents the knocking of brick rubble into the cavity and keeps the wall structurally sound. After removal of the outside face of the brick, cut a queen closure  1³⁄₁₆"(30 mm) thick. Now tip both ends of the brick and run the trowel around the top arris of the brick. Mortar up the remains of the brick in the wall and push the closure into place making sure the mortar compacts, filling the void between the two cut bricks.

It is sometimes necessary to hold the brick in place for a short period to allow the mortar to dry, then plug the mortar with the toothing filler.

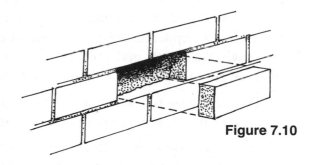

**Figure 7.10**

Cutting out chipped or damaged bricks that form a corner edge of a pier can be tricky. The best bet is to try not to get the entire brick out. Cut out the bed joint with a power saw fitted with a masonry blade. Chisel out the perpend joints with a plugging chisel and then use a bolster to remove enough of the old brick to be able to fit a new face brick.

The new face brick is just a cosmetic upgrade. It can't replace the bond nor the place of the original brick.

Use the diamond saw to cut the brick to be inserted to the size and shape required and use plenty of mortar to fit it.

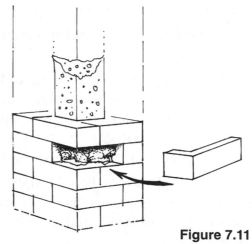

**Figure 7.11**

# 8
# Damp Proof Courses

## Introduction

The fitting of damp proof courses (DPCs) in a cavity brick wall is the job of the bricklayer, and if it is left out it will bring a multitude of problems for the owner of the structure.

Depending on the specifications of the foundation plan and the layout of the site, the bricklayer should lay a screeding course of bricks when laying the base of a double-skin brick house (and on some single-skin constructions).

This is a course of "common" or second-grade bricks that can be removed directly after the concrete floor has been poured. The screed brick should be buttered together without mortar in the perpends and the backs painted with sump oil to allow easy removal of the bricks, away from the freshly poured concrete floor.

The course should be stepped in the required distance to allow for the cavity—generally 1¾" or 2" (40 or 50 mm) (refer to Fig. 8.1).

Having removed the screed course, a rebate of one full brick in the floor will be left on, to which the damp course material can be fixed.

This method omits the cutting of splits of the first course on the inside skin, which occurs when the concrete floor is raised as in the brick veneer situation.

Fold the DPC, instead of cutting and joining.

When laying the first course of bricks on the DPC start with a weep hole, and mark the weep holes before laying to keep even distances, especially when there is more than one bricklayer working on the wall.

The DPC has to be laid higher than natural ground level, and has to be stepped accordingly to accommodate unlevel building sites.

**Figure 8.1**

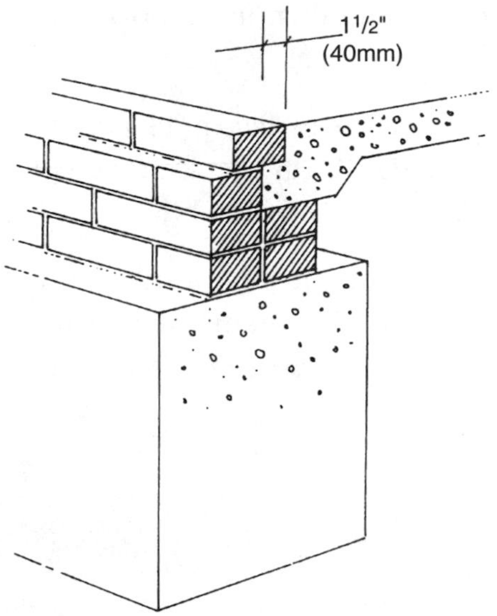

1½"
(40mm)

Where it has to be fitted above the roof by the bricklayer, allow plenty of lap in the DPC and also make sure it is high enough not to come out below external flashing.

The correct fitting of damp proofing is vital for the prevention of water problems. In the fitting of DPCs over the tops of doors and windows, slit the building paper and tuck the damp course up and in behind the paper.

Refix with buttons, turning the damp proof course into the perpends. This is most important as it stops the water from moving along the brick courses. (See Fig. 8.4.) (Refer to the section on "Sills" for the fitting of a DPC around veneer windows Page 62.)

DPC over, under, and at the sides of windows and doors in a brick veneer house should also protrude past the window or door, enough to allow the DPC to be turned into the perpends.

Turning of the DPC has to be done as folding. Strictly, no cutting. To be folded, a more rigid DPC is to be used.

**Figure 8.2**

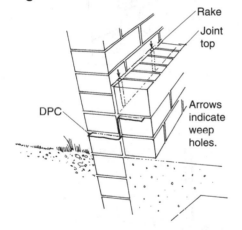

Waterproofing single-skin brickwork requires planning, especially for placement of DPC. When constructing the outside skin of brick, lay the DPC on the mortar bed ensuring 8" (200 mm) of DPC hangs inside the wall. (See Fig. 8.3.) This allows the DPC to be stepped up, one course to be laid under the header course of the inside "ledge" wall. The extra skin of brick inside creates a ledge. Moisture that penetrates the outside skin runs down the back of the wall onto the DPC then out weep holes poked in the perp joints along the DPC course. Rake the joint between the back of the wall and the header deeper so it acts as a catchment on the top of the headers so water will not run down the inside face of the header. Water penetration problems with single-skin brickwork usually occur in carports, external garages and under high-set houses. The same method detailed above and shown in Fig. 8.3 can also be used with cavity brickwork.

**Figure 8.3**

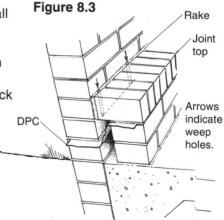

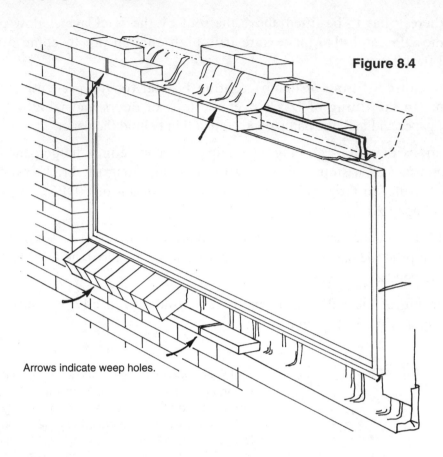

**Figure 8.4**

Arrows indicate weep holes.

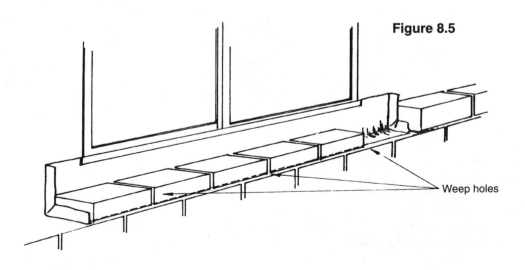

**Figure 8.5**

Weep holes

# Fitting DPC into Raked Ceilings

For raked ceilings choose a DPC with an internal aluminum lining, allowing the DPC to keep its preferred shape. When providing a damp course over raked ceilings there are some points to know:

- The exact rake of roof—this defines the height of the flashing to be stepped.

- The type of roof—whether it is galvanized iron or tile, and

- The type of external flashing to be used—lead or metal.

The DPC should be placed on the appropriate course, and extend out 100 mm from the face of the brickwork.

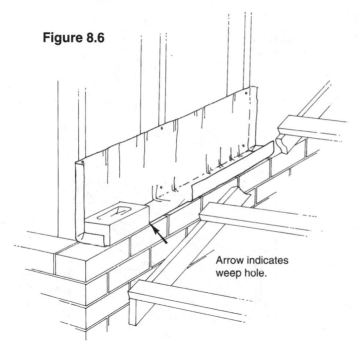

**Figure 8.6**

Arrow indicates weep hole.

Fit the flashing as in Fig. 8.6. This method is extremely efficient. It pays to always place more flashing than not enough, as the flashing is turned up on both ends and on the rafter side of the roof. We are not governed by any specific measurements, the flashing can go into the ceiling at any distance (preferably at the bottom of the top cord of the roof truss) and outside, allowing enough space for the batton iron or tiles for flashing and allowing at least a full brick to accommodate a weep hole. Again, flashing can be laid further out than normal as the excess can be trimmed off later.

**Figure 8.7**

Arrows indicate weep holes.

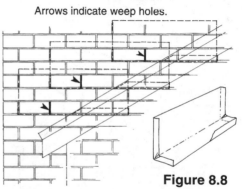

**Figure 8.8**

In this manner on completion of the brickwork any DPC in the ceiling will be tacked vertically up on to the face of the brickwork.

**Figure 8.9**

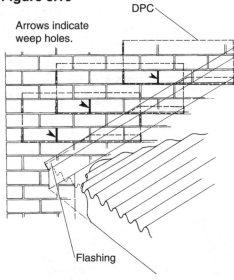

Weep holes will be left out as indicated. The excess DPC can be trimmed off after the flashing has been fixed to the brickwork.

Arrows indicate weep holes.

Trim off excess flashing with a Stanley knife.

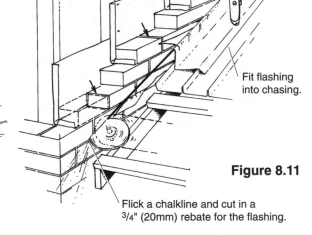

Fit flashing into chasing.

**Figure 8.11**

Flick a chalkline and cut in a ³/₄" (20mm) rebate for the flashing.

**Figure 8.10**

Arrows indicate weep holes.

DPC

Flashing

This method of laying the flashing is very efficient and can be placed easily. I think the finished result also looks much better than stepped flashing. It only takes a little time and effort in the planning.

# Fitting DPC into Single-Skin and Brick Veneer

Many local authorities are now introducing new requirements for the placement of damp proof coursing. The DPC, as was mentioned, stops rising damp. It is also recommended in the building of single skin brickwork to prevent moisture being drawn up the wall producing unsightly mold growth, not to mention threatening the strength of the wall from being subjected to constant water saturation.

However, although solving this problem, fitting DPC has created another problem and that is maintaining the strength of the bond between the courses where the DPC is placed. Previous methods of placing damp course on bricks, mortaring over DPC, and bricking on top, left no vertical bond strength between the courses.

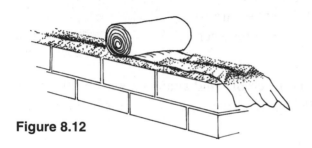

New methods recommend placing a bed of mortar on top of the brick course over which the damp course is laid, following which another bed of mortar is laid before the next course of bricks is laid.

**Figure 8.12**

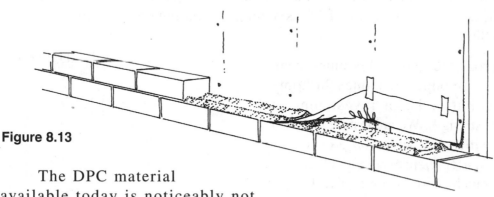

**Figure 8.13**

The DPC material available today is noticeably not smooth. The rough texture allows adherence to the mortar and therefore creates a bond between the courses where it is fitted.

Where a brick wall is being erected as an outside skin to a framed-up wall, or in the case of a cavity brick wall, the best method I have found is to fit the DPC behind the building paper and then curl it back and tape it to the inside wall before laying the mortar bed.

It is not advisable to nail it back because the holes created by the nails defeat the purpose of damp proof coursing.

Lay the mortar bed and splay the front edge. Fit the DPC over the bed, then run your trowel along the front edge so the DPC will follow the same splayed edge of the mortar bed over the DPC and continue bricking.

The integrity of the structural soundness of a brick wall is important, particularly in areas of high wind and cyclones. If anywhere along the DPC you have to start a new roll or join the DPC, make sure to lap it about 12" (300 mm) and to place mortar between the lap joint.

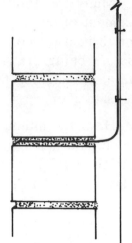

**Figure 8.14**

The bond strength should resist the force of wind against a wall. In the early days the courses that carried the DPC were slip jointed and didn't offer great wind resistance.

In our work we have found that the new local authority requirements for mortar beds top and bottom of DPC is adding several hundred dollars to the cost of a dwelling.

Across the nation this must mean enormous amounts of money in home building but it is an investment in structural strength. We are testing our own method which has yet to be proved—or for that matter disproved—as effective, but it is much more cost-effective for the home owner.

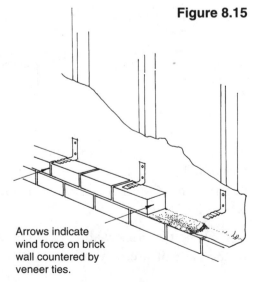

**Figure 8.15**

Arrows indicate wind force on brick wall countered by veneer ties.

We have tested and proven a process whereby we still use a slip joint, that is we place the DPC on the brick course, mortar on the DPC, and lay bricks on top, but on each stud on the internal wall of a timber frame dwelling we nail a veneer tie to the top of the course above the damp course to tie the wall in. This is counteracting the slip joint.

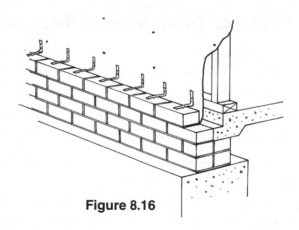

**Figure 8.16**

This method could be improved by using a rebate of one full brick, explained on page 118. Veneer ties could then be nailed to the bottom plate at intervals of one brick or less. In a cavity brick house we fit a cavity tie between the skins and top of the course above the DPC one and a half bricks apart.

Fitting DPC requires attention to detail. Ill-fitting joins or not returning the DPC around corners defeats its purpose, that is to stop rising damp, to stop water penetrating inside.

Pay particular attention to corners. Make sure the DPC is fitted behind the building paper.

When you reach a corner, fold the DPC around the corner (shown in Fig. 8.17) and allow a 12" (300 mm) lap. Work from the other side back to the corner to complete the lap. Seal any joints or holes with silicone.

**Figure 8.17**

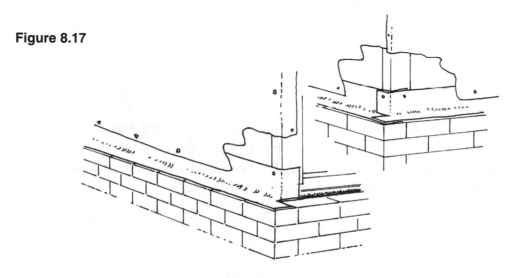

# Fitting DPC over Windows and Doors

Mostly 4" × 4" (100 × 100 mm) galvanized lintels are fitted over doorways and window openings. This creates some problems because the lintel sits some ⅜" (10 mm) from the face of the brickwork so it fits flush with the raked-out joint.

It is not advisable to fit it back too far because it can expose the holes in the underside of extruded bricks laid on top of it.

We prefer to set it back ⅜" (10 mm) and to slice 1" (25 mm) off the back of each brick that is laid on the lintel. The DPC is then fitted up under the building paper and fastened. We prefer to bring the DPC down over the first course of bricks above the lintel beam rather than have to fit it in behind bricks on lintel.

**Figure 8.18**

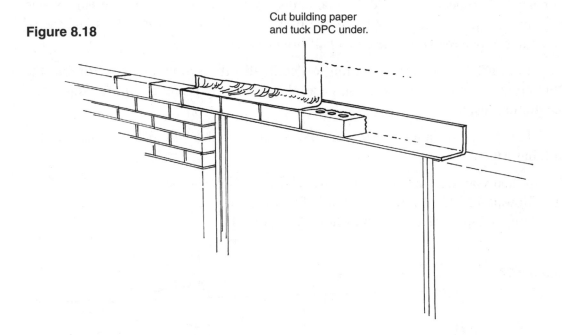

Cut building paper and tuck DPC under.

Because the lintel beam extends up 1" (25 mm) past the first brick, it eliminates any problems with water coming through the brickwork on the lintel beam under the DPC. On top of the DPC, weep holes are left in every fourth perpend. Make sure you prop the opening to prevent sagging.

# Replacing Damp Course in Brickwork

This method can also be used to place the DPC where it has been lifted out or where it is required to be added, for example, where an extension has been put on and another roof adjoins, requiring flashing.

**Figure 8.19**

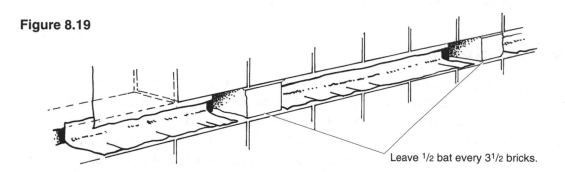

Leave 1/2 bat every 31/2 bricks.

Cut out three and a half bricks at a time leaving a half a brick in-between for structural support. Power tools are the best for the brick cutting because as little as possible hammering and banging is desirable on the supporting brickwork. I have an old saying that goes, "more cutting less cracking." Place the DPC and make sure it fits up under the paper to fit three bricks. The next day when the refitted mortar is dry remove the half brick that was left out for support the previous day and fit the DPC here and replace the brick. Be sure to use a bead of silicon on the DPC where it laps. Take care to look at the structural side of the building. I have successfully completed this method with six stories of brickwork on top.

**Figure 8.20**

Be aware that some products contain sugar substance that attracts ants and other vermin. This explains why these pests are often found on kitchen cabinets and in bathrooms.

Sykaflex all laps.

On this work I prefer to use M3 mortar as it dries fast, has a stronger MPA, and is better for the job because we want to remove the half bricks, as soon as possible for the obvious reason that we can never predict what the weather will do or how the wall loading will be affected. When fitting these bricks, refer to page 115 for details on how to best fit the bricks using wedge bed joints. Also be sure to round iron the joint, packing the bed in suitably.

# Weep Holes

Too many or too large weep holes can cause a lot of trouble. In heavy weather conditions, driving rain backed by strong wind actually pushes water into the weep holes if they are too large. Strong wind can then pressurize the openings and back the water up in the cavity. The water builds up as the wind prevents escape. I would be happy with a maximum of 1" high × ⅜" wide (25 × 10 mm) weep holes.

# 9
# Blocklaying

## Introduction

We talked earlier about the different types of house construction. Block houses are of masonry construction but utilize a concrete or clay block of larger dimension than a normal brick. These houses have great strength and are simple to erect and maintain.

Clay blocks are relatively new to the building game. They have eventuated over the last ten years, which is new in building terms, when you consider the Romans were making concrete blocks thousands of years ago.

Clay blocks are economical to use because they are double faced which means one block is used for the finish for the inside and outside of the wall. Clay blocks remove the need for timber framing, veneer, or wall sheeting during the construction process. They require no painting, are light, and are structurally very strong, which simplifies the construction process. Clay blocks may need to be sealed to manufacturer's specifications.

Clay blocks are often used in rental properties and in high wind areas because of their structural properties and the fact that they are a virtually maintenance-free building product. Recent tests have also found they have desirable qualities for construction in regions prone to earthquakes.

Clay blocks are the same height as a brick [3⅜" (86 mm)] but they are 12" (300 mm) long and 6" (150 mm) wide. They give the look of a brick veneer house and construction is not too different from concrete masonry block construction.

The first two courses will need to be E blocks to contain the concrete slab, depending on engineering requirements. To avoid rising damp, the DPC is laid over the last block below floor height and returned up the inside of the next course to be trimmed later.

Now it is time to look at the methods associated with the construction of these dwellings or buildings. The learner will quickly come to understand that there are some variations in blocks.

The commonly available block sizes are 3⅝" (90 mm) wide, 5½" (140 mm) wide and 7⅝" (190 mm) wide. Blocks are usually 15⅝" (390 mm) long and 7⅝" (190 mm) high. This allows the easy addition of ⅜" (10 mm) to round off the block sizes for estimating. The ⅜" is our mortar bed. That gives us blocks of varying width but a finished stand of 16" (400 mm) long and 8" (200 mm) high, for working out our gauges and lengths for fitting in and maintaining consistent modules.

The 3⅝" (90 mm) blocks are normally used internally and are rarely used as outside walls. The designer of the building will more often than not specify the width of block required to comply with structural strength needed to withstand pressures brought about by weight and weather. The 8" (200 mm) blocks are mostly used for external walls because of their thickness and strength. It is also likely that hollow block walls will need to be filled around doorways and windows and around the top bond-beam, using concrete to give additional strength.

There are split-face blocks available which have a rough-cast external face giving a more rustic look and getting away from the flat look in walls associated with laying hundreds upon hundreds of standard square-face blocks.

New style clay blocks are 12" (300 mm) long (with mortar) and the same height as a brick, 3⅝" (90 mm) including the mortar. They are a brick look-a-like about 5½" (140 mm) wide but they have a structural variance to bricks that puts them in the same category as blocks regarding wind and structural bearing.

There is a top and bottom side to each block. The top side is thicker all around the web for the obvious reason of having mortar spread on it. Sometimes it becomes necessary to butter the other side of the block which has a thin edge or web. This is most common when blocking over doorways and windows, when the blocks are used upside down and cut out to accommodate the laying of reinforcing steel rods. These blocks are called "knock-out blocks."

Unlike bricks, blocks are sold in very accommodating sizes. The estimator can have supplied half, three-quarter, and quarter blocks, and even specially designed sill blocks. Does this ever save a lot of time in cutting!

The half blocks are used to start each second course to obtain our stretcher bond. With bricks we would need to cut a brick every second course. The three-quarter size blocks work very well with fitting up to windows not in modules of 2' (600 mm).

Again here in our descriptions we are describing the method for right-handed layers and those who operate with their left hand should simply carry out the movements vice versa.

In laying blocks, buttering both edges of each perpend can be completed in one move. This requires twice as much mortar on the trowel as is needed to butter a brick.

**Figure 9.1**

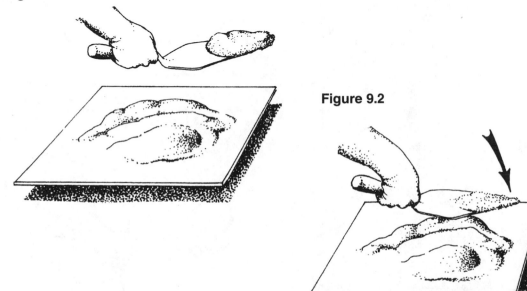

**Figure 9.2**

Run the trowel downwards in a rounding motion. When the mortar touches, it scrapes off onto the block. Move the trowel backwards, rotate the wrist clockwise, and tip the other side of the block as shown below.

Now using the bottom of the trowel, flatten the mortar on the inside of the perpends pushing down to the end of the block.

**Figure 9.3**

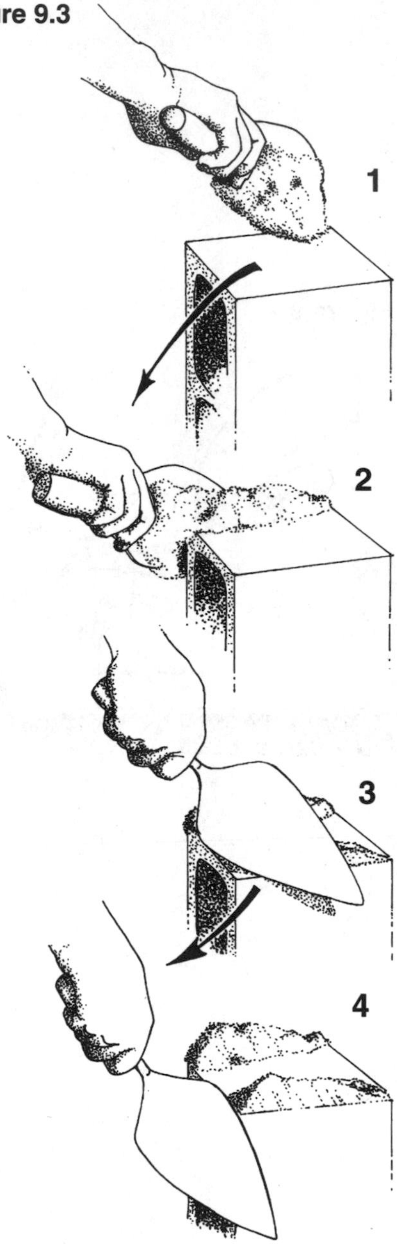

**Figure 9.4**

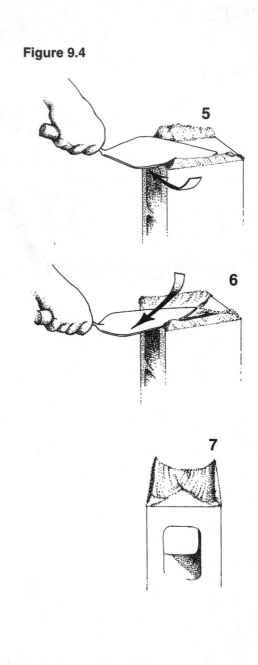

This will hold the mortar to the block while it is being laid, and make a fuller perpend. Buttering a block perpend in one move will save the extra movement associated with "double-dipping" for mortar. It is important to master this technique early in your career. It will save you one bend for every block you lay for the rest of your life.

Some layers prefer to butter the perpend with the block lying down. Grasp the block as shown, buttering the perpend the same as for a brick with the block in the left (or right) hand and the trowel working across the edges of the front perpend. Again pick up double the mortar required to butter a brick.

**Figure 9.5**

**Figure 9.6**

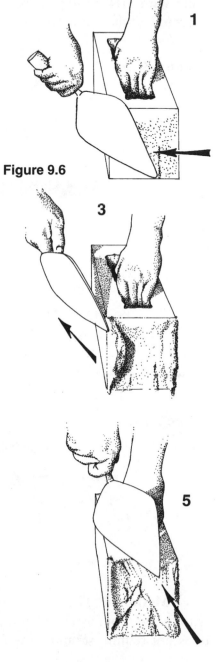

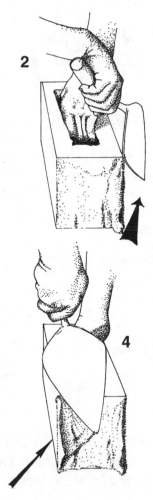

# Laying the Bed Joint for Blocks

For a mortar bed for blocks, scoop up as much as you can fit on your trowel. I always aim to pick up enough mortar to lay a bed the length of two blocks, which is approximately 32" (800 mm). It is compressed to the trowel in the same manner as is done with bricks—that is by dropping the wrist in a downward "flick," stopping short to bed the mortar to the trowel.

**Figure 9.7**

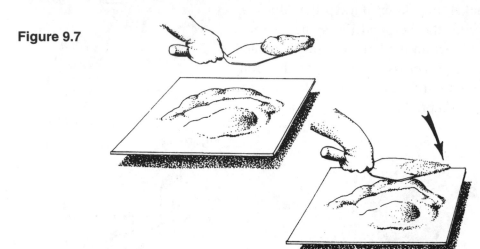

This movement fastens the mortar to your trowel and makes it easier when manipulating the trowel into positions to butter around obstructions in the bed, like protruding cyclone rods, and to lay a bed around the block when it has been laid upside down exposing the thin wall.

**Figure 9.8**          **Figure 9.9**

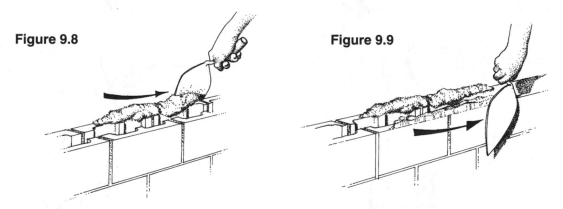

**Method 1**. The most commonly used method for laying out a bed for blocks is to throw it out as in the same way as throwing out a brick bed joint. This method does not require scraping the face of the trowel down over the edges and faces of

previously laid blocks. Any method that avoids this is worthwhile, because when laying split-face blocks this edge or face scraping will almost invariably lead to block faces being splattered with mortar.

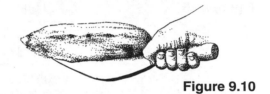

**Figure 9.10**

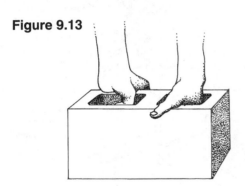

**Figure 9.11**

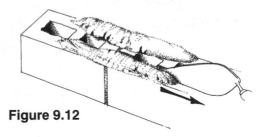

**Figure 9.12**

Take extreme care when working this way to avoid filling up the cores of the blocks with mortar. At the start it will seem almost unavoidable, and I'd personally like a dollar for every trowel of mortar that has fallen down the core of blocks before beginners have mastered the technique of laying a mortar bed for blocks. In the course of your laying career you avoid this happening as much as possible because of the waste of material and time.

# Laying Concrete Blocks

First, bring to the attention of the laborer that the mix for blocks has to be firmer than brick mortar —to support the block and to prevent the block from sinking below the line as they are heavy, especially the 8" (200 mm) blocks. This also helps prevent mortar slipping down the cores. Depending on the individual, blocks can be laid one or two handed. It depends too on the individual's size and strength, and the positions into which the blocks need to be maneuvered.

**Figure 9.13**

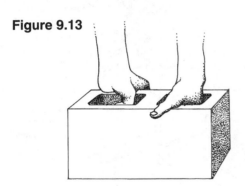

**Figure 9.14**

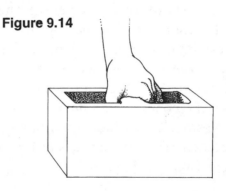

**Figure 9.15**

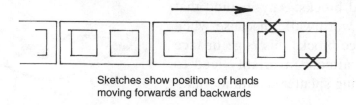

Sketches show positions of hands
moving forwards and backwards

**Figure 9.16**

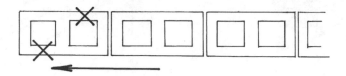

Blocks are pressed down in a manner similar to bricks, provided the mortar is the right consistency. When I'm working a block construction we mix our mortar the same as for brickwork except we replace the four shovels of loam with four shovels of a good, clean, fine sand. The type of raw material you will have to learn to live with will depend on your location in this world. Sands and loams vary according to geography.

Laying one handed will leave the other hand free to butter the block or tap it down to the line. Grasp the block by the middle core and tip the block as shown. The most important aspect to laying blocks with ease, speed, and the least effort, is the placing of the blocks, after spreading the bed and buttering the blocks.

Lift the block over the stringline, keeping it horizontally level lengthways and sighting down through the stringline. Place the block gently. Press the block to the stringline, adding only necessary finishing taps. True the block with the trowel while removing excess mortar. Remember to sight the perpends down.

Fig. 9.17 shows the use of knock-out blocks not only to retain the concrete floor but also the top of the block can be used as a level from which to screed the floor. Knock-out blocks are laid around the base of the building to allow the concrete and steel to travel from the floor to the footings without a requirement for boxing, a saving in time and money.

**Figure 9.17**

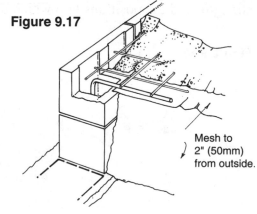

Mesh to
2" (50mm)
from outside.

Notice in Fig. 9.18 how the side of the block can be easily broken away allowing the concrete floor to pour into the block wall right to the footings in one go. Notice also the webs of the blocks are made to be easily broken away as in Fig. 9.19.

**Figure 9.18**

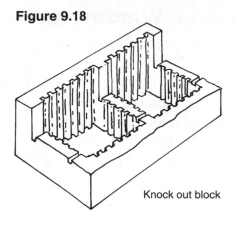

Knock out block

**Figure 9.19**

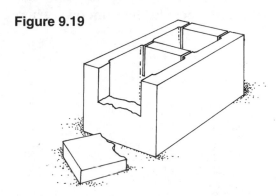

Knock-out corner blocks, as shown in Fig. 9.20 are also available. Reinforcing mesh can be laid right to the outside of the knock-out block which cannot be achieved by using E blocks as the ribs across the ends and the center prevent the fitting of mesh.

**Figure 9.20**

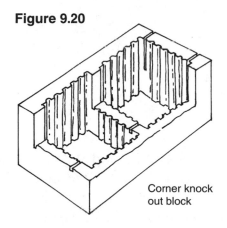

Corner knock out block

**Figure 9.21**

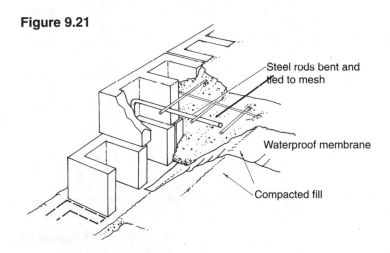

Steel rods bent and tied to mesh

Waterproof membrane

Compacted fill

# Block Window Sills

Depending on the circumstances, if blockwork has to be laid over the window (Fig. 9.22) then the top of the last course of blockwork will have to be ⅜" (10 mm) below the window. This allows the blockwork to sit on the window with a bed joint of ⅜" beneath the top of the window frame. When gauging the blockwork be sure to start the gauge ⅜" below the top of the window frame. If using lintel blocks, be sure to lay either knock-out blocks or lintels with holes beside the windows where vertical steel will have to be fitted.

**Figure 9.22**

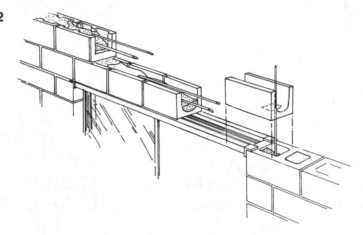

Figs. 9.23 and 9.25 show a window situation that requires the sill block to be laid higher. This pushes the window ⅜" above the last course and doesn't require us to grind the block gauge down to finish below the top of the window. Set-out of all walls should be ⅜" short of modular measurement and all openings ⅜" over. This is the first rule of setting out modular work. Frames should never be a problem.

**Figure 9.23**

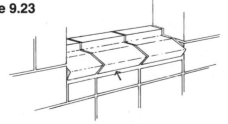

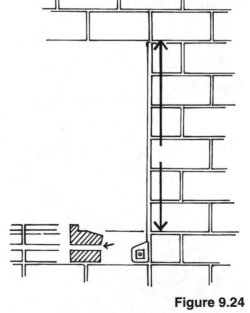

**Figure 9.24**

To ensure the sill is laid at the correct height, run a tape measure vertically up from the last block laid at correct sill height before the sill is laid. Allow the correct gauge above the blockwork to head height. Then mark the height of the window on the end of the block laid on top of sill height, remembering that if the blockwork has to be laid over the window, the last block has to finish ⅜" below the window.

**Figure 9.25**

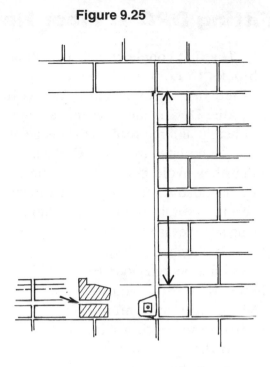

**Figure 9.26**

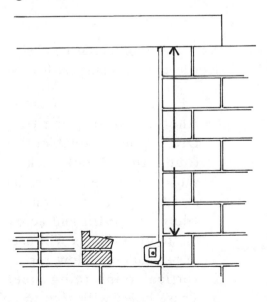

Fig. 9.26 shows how blockwork is laid when the window is required to be flush with the top of the last course of blockwork. Care is needed when laying the sill blocks. Sill blocks are generally made to work blockwork over the top and there is generally a difference of ⅜" or more, so after the sill is laid, make sure to gauge down from the top of the window to the first course of blockwork closest to sill height. Some manufacturers make sills to suit but standard windows are 4' (1200 mm) high, not 4'⅜" (1210 mm) as required.

# Fitting DPC at Floor Height or Below

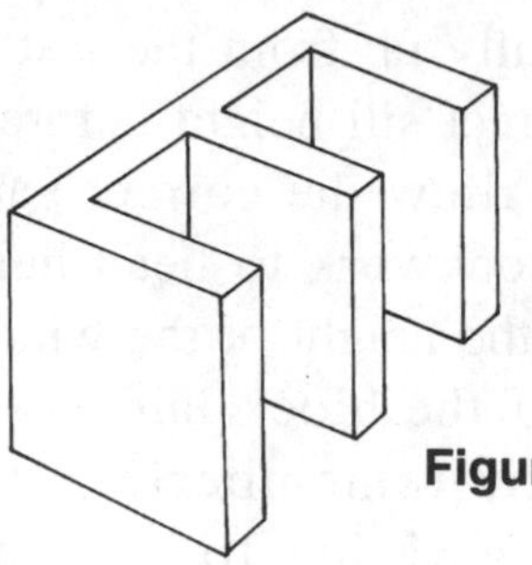

The first couple of courses will need to be E blocks to contain the slab, depending on engineering requirements. To avoid rising damp, the DPC is laid over the last block below floor height and returned up the inside of the next course to be later trimmed. E blocks are useful in that they avoid boxing up for the floor pour. They are also known as "floor blocks" as the concrete floor is poured into them tying the footing and the slab together. This is preferable to laying blocks on a poured floor. E blocks also allow you to bend the upright reinforcing from the footing and tie into the mesh for the floor which helps prevent movement and cracking of the block and concrete work.

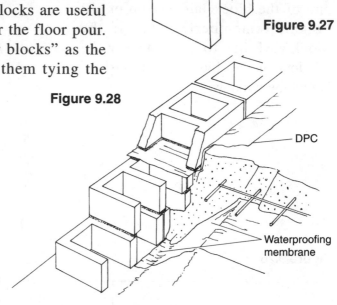

**Figure 9.27**

**Figure 9.28**

DPC

Waterproofing membrane

The DPC should be mortared in as in Fig. 9.28. This thin mortar bed below and above the DPC prevents the block course at floor level from being pushed out during the concrete pour. (See Figs. 8.12 and 8.13.)

## Preventing rising damp

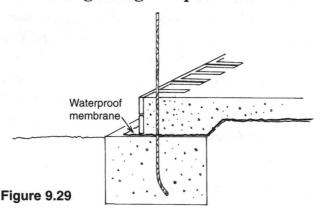

Waterproof membrane

**Figure 9.29**

Another option for fitting DPC when constructing foundation blockwork is shown in Fig. 9.29. Here the moisture-proof membrane is laid over the fill and across the footing. Holes will need to be made to allow for the vertical reinforcing steel. These holes will need to be sealed to prevent any moisture penetration into the concrete slab. After completion of blockwork, the waterproof membrane is turned up against the blockwork and retained by surrounding earth.

The course directly above the DPC is filled with concrete when pouring the floor. The following course will have to include E blocks where vertical steel is necessary. This allows you to clean out the cores where the steel is to be fitted. This is important as mortar falling down from 28 courses or more over a wall can almost fill the cores to the extent that vertical bars cannot be successfully fitted.

E blocks have a knock-out side, which is later replaced when the core is filled. Hosing is by far the best way to clean out the cores but timing is important because too much water too soon can severely wash out the mortar joints, weakening the wall, discoloring the mortar, and even discoloring the blockwork.

**Figure 9.30**

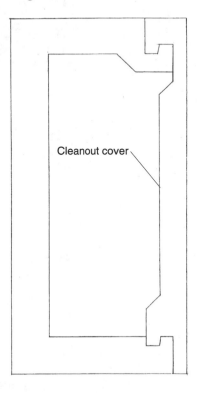

Cleanout cover

Figs. 9.30 and 9.31 show a clean-out block that I designed. It has the advantage of allowing the cover plates to be fitted at any time.

They can also be fitted at any course as their "face" will match existing work unlike the currently available clean-out blocks. The fitting of clean-out blocks allows the cover to be properly cleaned out prior to tying off the steel and filling with concrete.

**Figure 9.31**

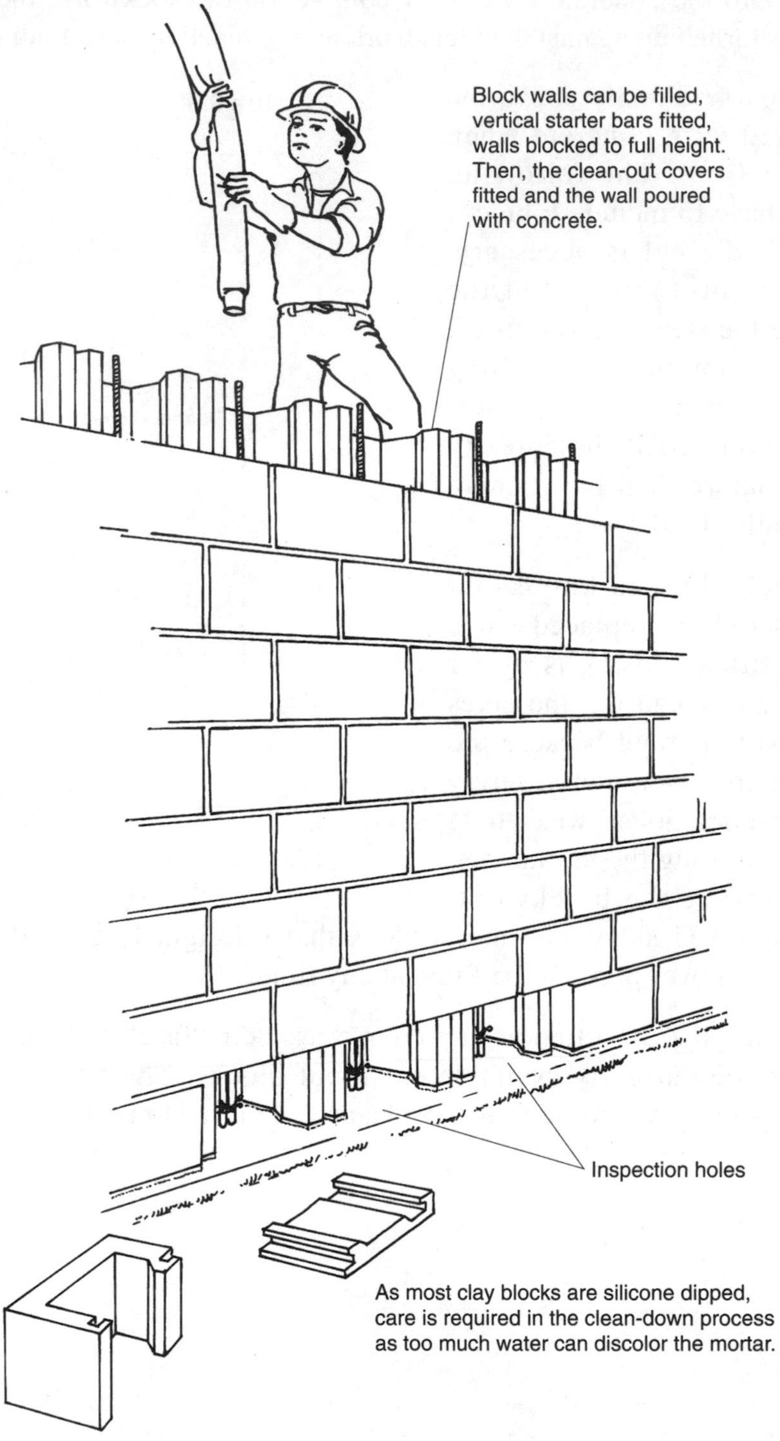

Block walls can be filled, vertical starter bars fitted, walls blocked to full height. Then, the clean-out covers fitted and the wall poured with concrete.

Inspection holes

As most clay blocks are silicone dipped, care is required in the clean-down process as too much water can discolor the mortar.

# Setting up Profiles on Clay or Concrete Blockwork

As I mentioned earlier I prefer to work from the outside for the simple reason that scaffolding and accessibility is so much easier. Run the stringlines and check all measurements as usual then set up the posts as I have suggested in Fig. 9.32. You can loop the line or use the line holders as shown. Now shoot a level on all of the posts and mark up the gauge. Sort out the bond that will be required and you are ready to start. For these constructions I use posts 1½" (40 mm), SHS (Square Hollow Section), 13' (4 m) long. These posts are there for the duration of the job and as they are 13' long, or full height, gauges are run from top to bottom and our brick work will then end up perfectly level and plumb, whereas shifting posts up or building corners can result in some variation.

**Figure 9.32**

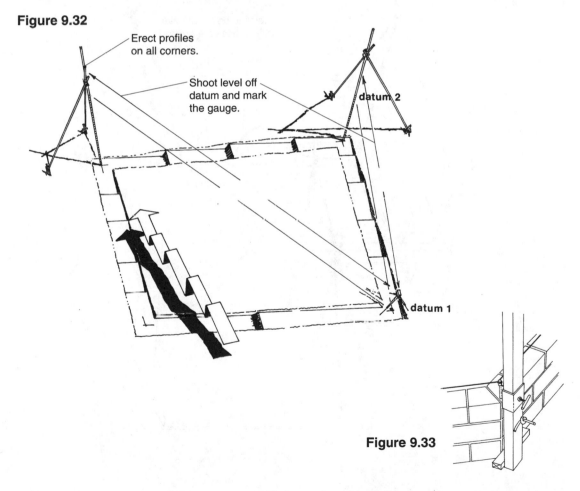

Erect profiles on all corners.

Shoot level off datum and mark the gauge.

datum 2

datum 1

**Figure 9.33**

# Working from the Inside

Fig. 9.34 shows working from the inside.

With split-face block, where the outside face of the E block can vary as much as ¾" (20 mm), the inside face of the block would look ghastly if laid from the outside. The side from which the blocks are laid will always be the best side, so for a family room or when a client wants the inside face of the block to look better, you will have to lay from the inside. On some sites there are bound to be problems erecting or fitting scaffolding, and those sites will require the brickwork to be laid off the floor on the inside.

Notice how the line is run around the posts and, in this case, is connected to the steel door frame by a steel corner block. For this setup we have G-clamped the window

**Figure 9.34**

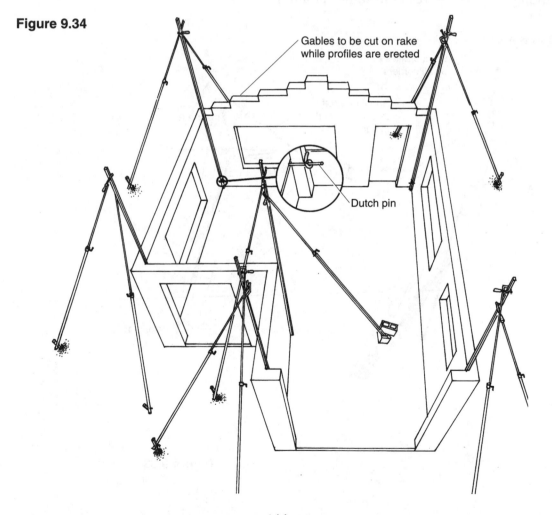

Gables to be cut on rake while profiles are erected

Dutch pin

props to the posts to hold them plumb. Fig. 9.35 shows the fitting of a profile into a doorway which does the same job as a deadman profile but clearly more effectively as the profile only has to be fitted once and after it is gauged, the line can be fitted and the brickwork completed.

When using split-face block (Fig. 9.36), the blockwork has to be laid from the inside because you can't get a good enough line on the rough exterior.

Fig. 9.37 shows fitting profiles to stop end for split-faced blocks. As in Figure 9.36, the bottom of the profile can be held by laying a piece of block on a small bed of mortar to hold the bottom of the profile firm. After the first course is laid, a sliding profile clamp can be fixed to the bottom of the first block with a sash cramp. With the profile in place, I then measure off the inside of the post to the first perpend on each course, thus keeping bond.

**Figure 9.35**

Deadman profile

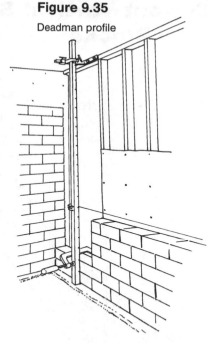

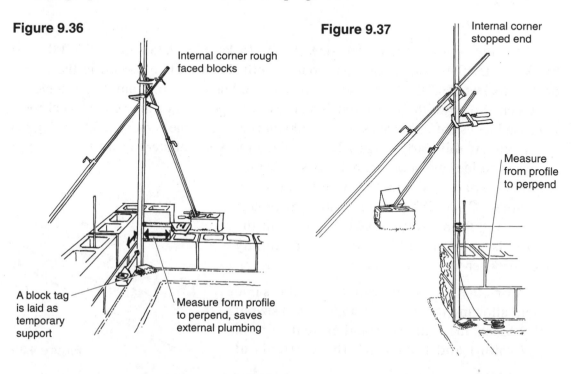

**Figure 9.36**

Internal corner rough faced blocks

A block tag is laid as temporary support

Measure form profile to perpend, saves external plumbing

**Figure 9.37**

Internal corner stopped end

Measure from profile to perpend

# Set out Vertical Steel Fittings in Concrete and Clay Blocks

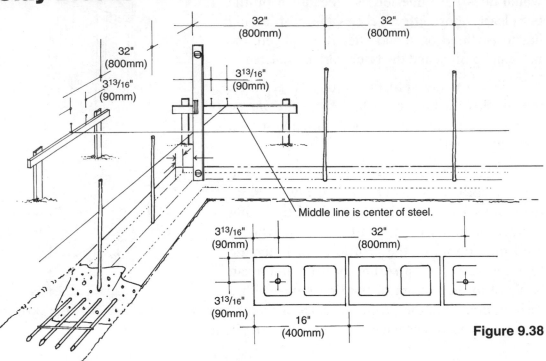

Middle line is center of steel.

**Figure 9.38**

Set-out will depend on the size of the blocks. Fig. 9.38 shows 8" (200 mm) blocks. After the footing has been poured, shift the stringline inward to the center of the blockwork. If the blocks are 8" we shift the line in 4" (100 mm) to the center of the block. Now run the tape through and mark the position of the vertical steel rods. Because blocks are in modules of 16" (400 mm) we mark the centers at 32" (800 mm) as shown. At this early stage I like to place the vertical steel at all the required positions besides the doors and windows and at the required spacings for the wind loadings, sometimes 32" or 48" (800 or 1200 mm) centers may be required. Vertical steel right from the footing is important to me, as it is structurally much stronger for construction.

For clay blocks we would use the same system allowing the 6"/12" (150/300 mm) size of a clay block, meaning we would move the line in 3" (75 mm) and then mark the verticals at

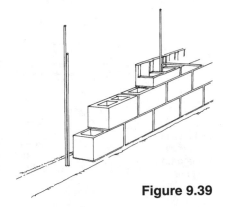

**Figure 9.39**

30"/36" (750 and 900 mm). Notice how I have two ½" rods in the footing, one rod continues to the top plate, the second rod is bent into the concrete slab.

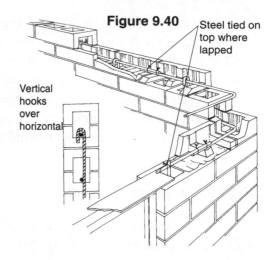

**Figure 9.40**

Steel tied on top where lapped

Vertical hooks over horizontal

# Fitting of Steel

Fitting of steel has to be done in conjunction with the bricklayer and the supervisor. All horizontal steel has to be laid during construction either by the bricklayer or by others. Be sure to position the steel correctly as specified in the plans. Make sure all corner bars are in and that all internal or butt walls are tied correctly with steel bent into both walls. It is far easier to use knock-out blocks where the steel sits on a bed joint to allow the concrete to pour around. When block widths are smaller or where there are more than two lengths of steel to tie, be sure to lay the steel one on top of the other and tie them as in Fig. 9.40. This allows access for concrete to be poured down beside the horizontal and vertical steel.

In some cases, there could be a number of lengths of steel reinforcing around which the concrete has to be poured. This situation is found in lintels above windows and doors, reinforcing that may be required under windows and in bond beams found at the top of walls (Fig. 9.42). These linkages are fitted to lintels usually greater than 1.7 m in high wind areas, giving the lintel lateral strength.

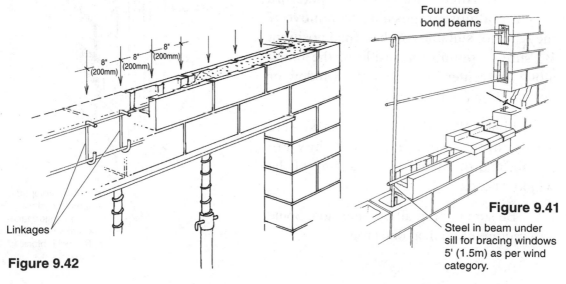

Linkages

**Figure 9.42**

Four course bond beams

**Figure 9.41**

Steel in beam under sill for bracing windows 5' (1.5m) as per wind category.

On occasion, the steel reinforcing will have to be linked together as the wall is constructed. This is most likely where the cores of the blockwork have to be filled. In all blockwork, time should be taken to position the starter bars correctly. Incorrect positioning will not only cause a headache for the bricklayer later, but can also cause severe bending of the rods and possibly result in structural defects. Steel thicker than ½" (12 mm) is virtually impossible to bend by hand so preliminary thought has to be given, especially for steel around windows and doors.

# Fitting Windows

Windows and doors are generally built into the blockwork by the bricklayer during construction of the wall. However, there are times when windows and doors need to be fitted after the blockwork is completed. When constructing openings for these situations, be sure that you carefully select the blocks to be used. Any blocks that are out of shape will cause troubles with the fitting of the windows and doors. You should not use chipped or damaged blocks around the perimeter of any opening. In most cases, the core of the blocks are filled with concrete, therefore they are almost impossible to remove if they are not suitable for their final position. Be sure to plumb the inside of the block along the center line where the window or door will be fitted.

Figure 9.43 shows how to make sure all openings are square, that they are plumb, and that there is no sagging of the window heads.

Be sure to prop all openings too. Some will require more than one prop.

**Figure 9.43**

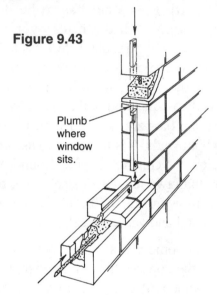

Plumb where window sits.

**Figure 9.44**

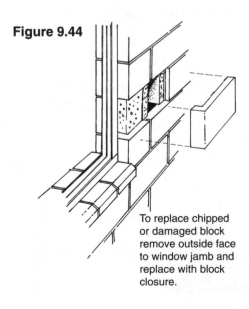

To replace chipped or damaged block remove outside face to window jamb and replace with block closure.

The astute bricklayer will be on the lookout for situations that have to be taken into account before blockwork is run up. Be on the lookout for different levels as in the case of patios (Fig. 9.45). In this situation, where the rooflines are lower, piers or walls could be lower, requiring a split brick on the top course. In commercial

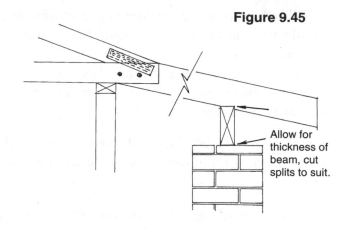

**Figure 9.45**

Allow for thickness of beam, cut splits to suit.

situations there are often thickening beams or supporting beams or suspended floors of different thickness, which have to be taken into account. All have to be allowed for by the bricklayer.

Other traps to be aware of with blockwork are where beams are placed in the last two courses. Read the plan thoroughly as there may be a need for cutouts for beams, service ducting, or the fitting of air conditioners. It's far easier to plan the blockwork than to cut through reinforcing steel and concrete after the blockwork has been completed.

# KO Beam over Openings

Cut a piece of ⅜" (8 mm) Hardiplank (or other fibrous cement sheeting) 2¼" (60 mm) longer than the length of the window so it will extend over 1⅛" (30 mm) each end when placed on top. Cut it width ways ⅜" less than the type of masonry construction—for 7⅜" (190 mm) blocks deduct ⅜" which gives 7¼" (180 mm).

Now lay the blockwork finishing ¹⁄₁₆" (2 mm) below the top of the window height. Place the Hardiplank with the smooth side down on top of the window, protruding 1¹³⁄₁₆" (30 mm) past each end of the window. The blocks can be laid directly on top of the plank.

We finish ¹⁄₁₆" (2 mm) below the window because our mortar bed is normally ⅜", and the ¹⁄₁₆" plus the ⁵⁄₁₆" (8 mm) thickness of the plank will equal a standard mortar bed and keep our courses even. Place vertical stays under the window. These will be removed after the bond beam has been filled and set.

The blocks above the doorways and windows are laid upside down, that is with the thin web upwards. "Knock-out" blocks are made for this purpose and the ends and centers of the blocks are tapped out. Lay the reinforcing rods along the length of the head, or—depending on the plans—continue the knock-outs and steel laying along the entire course that runs the perimeter of the construction on top of the window or door height.

This course is filled with concrete to bond these blocks together. Doing so allows them to stand alone above the windows or doors without crushing down on the frames.

When blocking up to windows or sliding glass doors, use "sash" blocks that have a vertical cut-out to accommodate the fin on the aluminium frame. Filling the block sash with mortar will hold the fin on the frame into the blockwork, and prevent any window-frame movement in windy conditions.

If the fin has been removed, mortar up the end of the frame. Wire ties are then used to hold the open end of the window or door into the blockwork.

**Figure 9.46**

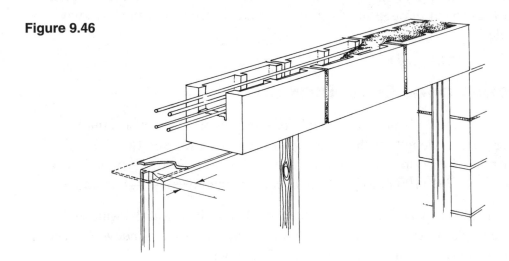

# Window and Door Heads of Clay and Concrete Block

The use of fibrous cement sheeting, cut to length and width, over the top of window frames is not only a time saver for the bricklayer but it makes obsolete all the cutting of timber bearers and extra propping. It also removes the need to finish off jointing underneath the window after the bond beam has been poured with concrete and the timber heads removed, because it is already neatly finished. We use lengths of Hardiplank which is commercially available in 20' (6 m) lengths and varying widths and thicknesses. I often use the ⁵⁄₁₆" (8 mm) thickness and I strip back the width to 7¼" (180 mm).

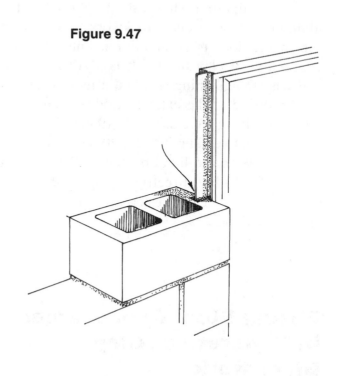

**Figure 9.47**

Figure 9.48 shows temporary window heads made of 1½" × 1½" (40 × 40 mm) steel square hollow section. An angle bracket is welded on each end. The square hollow steel section is actually curved upwards about 3 mm to allow for the weight of the blocks which are quite heavy. When fitting these temporary heads to windows over 3' (900 mm) wide, it is much easier to get them out if you let them protrude about ⅜" from the face of the block. Notice also the DPC laid across the cores that won't be filled. This saves using very heavy solid lintel blocks or trying to jam cement bags down the cores. The lintels I use are multipurpose with a knock out recess. This allows fitting of vertical steel down

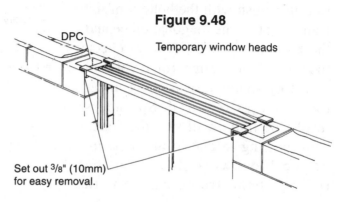

**Figure 9.48**

Temporary window heads

DPC

Set out ³⁄₈" (10mm) for easy removal.

the block beside each opening and makes filling the cores easy.

This diagram shows the bricking in of an aluminum door in clay blockwork. The plate on top of the door is an aluminum lintel designed especially for clay blocks. It is ¼" (7 mm) thick and has strengthening ribs and a molded drip on the underside of the exterior head. Depending on the gauge, if we were using ¹⁹⁄₃₂" (9 mm) bed joints, I would aim to end up ¹⁄₁₆" (2 mm) below the top of the door. With the lintel on top of the door, the blocks would lay directly on top of the lintel. Usually these lintels stick across ¾" (20 mm) each side of the door.

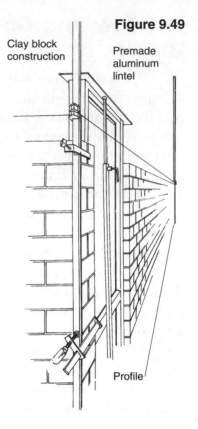

**Figure 9.49**

Clay block construction

Premade aluminum lintel

Profile

# Fitting Steel Door Frames in Concrete or Clay Blockwork

Blocks are laid from the inside after the door frame is plumbed and leveled and set at the exact height to allow blockwork to sit neatly on top of the frame and the concrete floor to finish flush with the bottom of the frame. Notice the diagonal brace and the two other braces. Always check the top of the frame for level and adjust by wedging or packing under the frame as required. Spreaders inside the frame will stop the frame from bowing in due to the pressure of the blockwork and from the filling of the frame cavity with

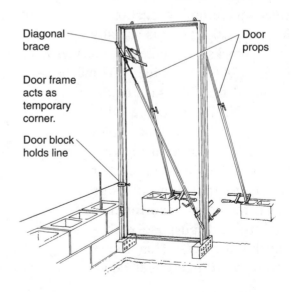

**Figure 9.50**

Diagonal brace

Door props

Door frame acts as temporary corner.

Door block holds line

mortar. When the frame is blocked in, it is difficult to make adjustments and therefore ensuring it is plumb, level, and square as work progresses is vitally important.

In Fig. 9.51, the door fitted to this frame swings internally. Notice the margin between the hinge and the blockwork—⅜" (10 mm). This is the maximum that will allow the door to open fully back against the wall. Locating the frame flush with the inside of the blockwork is also acceptable.

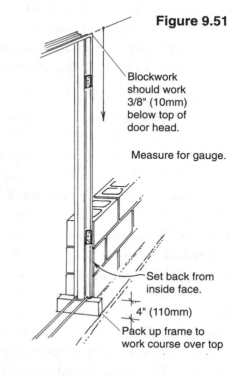

**Figure 9.51**

Blockwork should work 3/8" (10mm) below top of door head.

Measure for gauge.

Set back from inside face.

4" (110mm)

Pack up frame to work course over top

As I said before, it's the small things that make for a better finish and that take the headache out of brick and blockwork. For both of the frames in these illustrations I have placed a brick on edge to pack it up 4 ⅜" (110 mm) not 4" (100 mm) as expected. This additional ⅜" is required because the standard height of door frames is 7' (2.1 m).

Therefore adding 4⅜" to 7' we end up with 7' 4⅜" and, as we are working in modules of 8", the frames will end up ⅜" above the course of block allowing the next course to be laid directly on top of the frame as can be seen in Fig. 9.52.

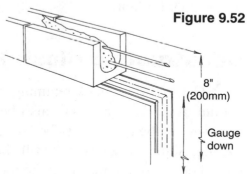

**Figure 9.52**

8" (200mm)

Gauge down

When a door frame sits on top of the floor and blocks are to be laid on the floor, the trick is to mark the gauge up so the last course of blocks will end up 3⅝" (90 mm) above the door. This is because a door frame is 7' but blocks work 7' 4". Therefore, finishing 3⅝" above, allows two alternatives—a half height course of 3⅝" blocks or a ¾ lintel block [11⅝" (290 mm)] will work up to the next course.

I like to place a dry ¾ lintel on the frame and mark the gauge down from it. Because frames are nearly all 3' wide the blocks across the top should be made up of a ¾ block if the courses are horizontal or 4" (100 mm) ¾ lintels if the courses are laid vertically.

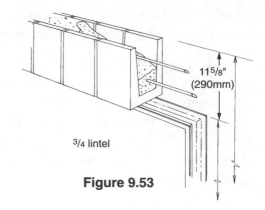

11⁵/₈" (290mm)

3/4 lintel

**Figure 9.53**

When fitting the frames have a look at the plan. If the length of the wall is an odd number, say, 40' 4" (12.1 m), and the frame is 3' wide, deduct this from the total length leaving 37' 4" (11.2 m), which will work blockwork without the need for ¾ blocks. Otherwise you will have ¾ blocks on both sides of the frame instead of only over the top of the frame. The ¾ blocks also make it hard to fill the cores containing the steel reinforcing as they lay across the core—most doors and windows have steel either side. Filling the frame cavity with mortar as you lay each block eliminates the need to tap the frame to make the mortar fill the cavity evenly and will avoid the possibility of the frame being knocked out of plumb. Spreaders should be used in the frame to prevent the frame from bowing and don't forget the ties to hold the frame securely into the blockwork. Ties should be placed evenly up the frame with care taken to ensure that the bottom of the frame is held in position. Extra ties should be used at the hinge position and either side of the keeper for the door furniture.

# Constructing Block Retaining Walls

Constructing block retaining walls that will withstand lateral pressure from behind or in front and will also be load bearing requires some skill. This wall however can be designed to be built anywhere. Notice the toe on the bottom of the footing. Some are towards the filled area, which allows for the compacted fill to be placed on top of the toe. With the footing this way the wall can be built much closer to another building or on a site without a great deal of access. On walls with compacted fill against them, especially in basements or garages, I leave the "snots" or "dags" around the mortar joints in place. I then mix a three to one mortar slurry mix and coat the side against the wall using a sponge. This has the effect of creating a bond across the face of the wall integrated with the mortar bed, which gives added strength. The common term is bagged brickwork and it's important it's done when the joints are wet so it bonds into the mortar joint effectively.

After the wall is completed and the mortar is dry I paint it with a waterproof paint to seal it and I then fit a moisture proof barrier of plastic. I like to stand ply against the plastic as it protects the paint and it protects the plastic from being damaged by rocks within the back fill. Because it's not possible to leave weep holes it's important to have plenty of drainage and this can be achieved with quality slotted drainage pipe. The slotted drainage pipe should be covered with about 12" (300 mm) of ¾" (20 mm) aggregate, which allows water to flow through easiest. On top this 12" of aggregate we back fill with ½" (12 mm) aggregate.

Engineering standards and load pressures will only allow 8" blocks to retain to a certain height. The amount of vertical and horizontal steel differs for each circumstance. I have included a table that lists the size and the amount of steel to use for various wall heights.

I can't stress enough, however, how important it is to read the plan. Some walls, especially those that support a concrete floor, are designed only for full strength after the floor has been poured.

The floor sits on the wall providing a lateral tie. This requires the wall to be propped prior to backfilling. My illustrations show footings with toes front or rear, but trench footings work equally as well. Be sure to know the recommended requirement for the width and depth of the footing for the height of the wall you intend to construct.

**Refer to Figs. 9.54 and 9.55 next page. Many thanks to Pioneer for the use of these tables.**

# Wall Type I

**Figure 9.54**

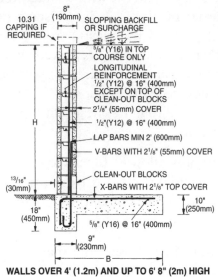

10.31 CAPPING IF REQUIRED
8" (190mm)
SLOPPING BACKFILL OR SURCHARGE
5/8" (Y16) IN TOP COURSE ONLY
LONGITUDINAL REINFORCEMENT 1/2" (Y12) @ 16" (400mm) EXCEPT ON TOP OF CLEAN-OUT BLOCKS
2 1/8" (55mm) COVER
1/2" (Y12) @ 16" (400mm)
LAP BARS MIN 2' (600mm)
V-BARS WITH 2 1/8" (55mm) COVER
CLEAN-OUT BLOCKS
X-BARS WITH 2 1/8" TOP COVER
13/16" (30mm)
18" (450mm)
9" (230mm)
10" (250mm)
5/8" (Y16) @ 16" (400mm)
H
B

**WALLS OVER 4' (1.2m) AND UP TO 6' 8" (2m) HIGH**

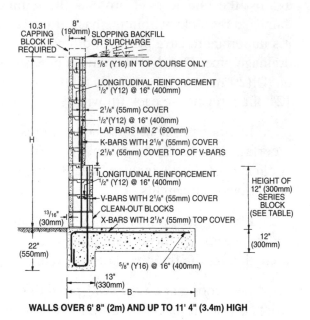

10.31 CAPPING BLOCK IF REQUIRED
8" (190mm)
SLOPPING BACKFILL OR SURCHARGE
5/8" (Y16) IN TOP COURSE ONLY
LONGITUDINAL REINFORCEMENT 1/2" (Y12) @ 16" (400mm)
2 1/8" (55mm) COVER
1/2" (Y12) @ 16" (400mm)
LAP BARS MIN 2' (600mm)
K-BARS WITH 2 1/8" (55mm) COVER
2 1/8" (55mm) COVER TOP OF V-BARS
LONGITUDINAL REINFORCEMENT 1/2" (Y12) @ 16" (400mm)
V-BARS WITH 2 1/8" (55mm) COVER
CLEAN-OUT BLOCKS
X-BARS WITH 2 1/8" (55mm) TOP COVER
13/16" (30mm)
22" (550mm)
13" (330mm)
5/8" (Y16) @ 16" (400mm)
HEIGHT OF 12" (300mm) SERIES BLOCK (SEE TABLE)
12" (300mm)
H
B

**WALLS OVER 6' 8" (2m) AND UP TO 11' 4" (3.4m) HIGH**

## BASE SIZES FOR WALL TYPE 1

| WALL HEIGHT "H" | BASE WIDTH "B" | | |
|---|---|---|---|
| | BACKFILL | | |
| | LEVEL WITH SURCHARGE | LEVEL WITH 5kPa* SURCHARGE | SLOPE UP TO 1 IN 4 WITH NO SURCHARGE |
| 2' 8" (800mm) | 2' (600mm) | 2' 8" (800mm) | 2' 4" (700mm) |
| 3' 4" (1000mm) | 2' 4" (700mm) | 3' (900mm) | 2' 8" (800mm) |
| 4' (1200mm) | 2' 8" (800mm) | 3' 4" (1000mm) | 3' 4" (1000mm) |
| 4' 8" (1400mm) | 3' (900mm) | 3' 8" (1100mm) | 4' (1200mm) |
| 5' 4" (1600mm) | 3' 4" (1000mm) | 4' (1200mm) | 4' 8" (1400mm) |
| 6' (1800mm) | 3' 8" (1100mm) | 4' 8" (1400mm) | 5' 4" (1600mm) |
| 6' 8" (2000mm) | 4' (1200mm) | 5' 4" (1600mm) | 6' (1800mm) |
| 7' 4" (2200mm) | 4' 8" (1400mm) | 6' (1800mm) | 7' (2100mm) |
| 8' (2400mm) | 5' (1500mm) | 6' 8" (2000mm) | 7' 8" (2300mm) |
| 8' 8" (2600mm) | 5' 8" (1700mm) | 7' (2100mm) | 8' 4" (2500mm) |
| 9' 4" (2800mm) | 6' (1800mm) | 7' 8" (2200mm) | 9' 4" (2800mm) |
| 10' (3000mm) | 6' 8" (2000mm) | 8' (2400mm) | 0' (3000mm) |
| 10' 8" (3200mm) | 7' (2100mm) | 8' 8" (2600mm) | 11' (3300mm) |
| 11' 4" (3400mm) | 7' 8" (2300mm) | 9' 4" (2800mm) | 12' (3600mm) |

*Note: 12" (300mm) of soil or a private driveway imposes a load of approximately 5kPa.

## REINFORCEMENT DETAILS WALL TYPE 1

| WALL HEIGHT | | | | REINFORCEMENT | |
|---|---|---|---|---|---|
| TOTAL HEIGHT "H" | HEIGHT OF 150 SERIES BLOCK | HEIGHT OF 200 SERIES BLOCK | HEIGHT OF 300 SERIES BLOCK | X-BARS AND V-BARS | K-BARS |
| 2' 8" (800mm) | 2' 8" (800mm) | — | — | 1/2" (Y12) | — |
| 3' 4" (1000mm) | 3' 4" (1000mm) | — | — | 1/2" (Y12) | — |
| 4' (1200mm) | 4' (1200mm) | — | — | 1/2" (Y12) | — |
| 4' 8" (1400mm) | — | 4' 8" (1400mm) | — | 5/8" (Y16) | — |
| 5' 4" (1600mm) | — | 5' 4" (1600mm) | — | 5/8" (Y16) | — |
| 6' (1800mm) | — | 6' (1800mm) | — | 5/8" (Y16) | — |
| 6' 8" (2000mm) | — | 6' 8" (2000mm) | — | 3/4" (Y20) | — |
| 7' 4" (2200mm) | — | 4' 8" (1400mm) | 2' 8" (800mm) | 5/8" (Y16) | 5/8" (Y16) |
| 8' (2400mm) | — | 5' 4" (1600mm) | 2' 8" (800mm) | 5/8" (Y16) | 5/8" (Y16) |
| 8' 8" (2600mm) | — | 5' 4" (1600mm) | 3' 4" (1000mm) | 3/4" (Y20) | 3/4" (Y20) |
| 9' 4" (2800mm) | — | 6' (1800mm) | 3' 4" (1000mm) | 5/8" (Y16) @ 8" (200mm) | 5/8" (Y16) @ 8" (200mm) |
| 10' (3000mm) | — | 6' 8" (2000mm) | 3' 4" (1000mm) | 5/8" (Y16) @ 8" (200mm) | 5/8" (Y16) @ 8" (200mm) |
| 10' 8" (3400mm) | — | 6' 8" (2000mm) | 4' (1200mm) | 3/4" (Y20) @ 8" (200mm) | 5/8" (Y16) @ 8" (200mm) |
| 11' 4" (3400mm) | — | 6' 8" (2000mm) | 4' 8" (1400mm) | 3/4" (Y20) @ 8" (200mm) | 5/8" (Y16) @ 8" (200mm) |

**ALL BARS ARE TO BE 400 CRS UNLESS OTHERWISE NOTED.**

# Wall Type 2

**Figure 9.55**

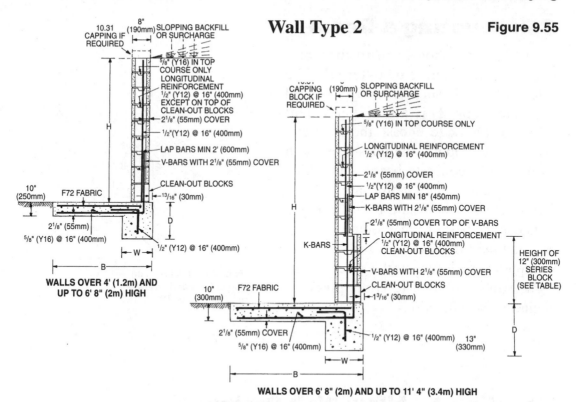

**WALLS OVER 4' (1.2m) AND UP TO 6' 8" (2m) HIGH**

**WALLS OVER 6' 8" (2m) AND UP TO 11' 4" (3.4m) HIGH**

## BASE SIZES FOR WALL TYPE 2

| WALL HEIGHT "H" | BASE WIDTH "B" | BASE WIDTH "W" | BASE DEPTH "D" | | |
|---|---|---|---|---|---|
| | | | LEVEL WITH NO SURCHARGE | LEVEL WITH 5kPa* SURCHARGE | SLOPE UP TO 1 IN 4 WITH NO SURCHARGE |
| 2' 8" (800mm) | 2' (600mm) | 1' (300mm) | 1' 8" (500mm) | 1' 8" (500mm) | 1' 8" (500mm) |
| 3' 8" (1000mm) | 2' 4" (700mm) | 1' (300mm) | 1' 8" (500mm) | 1' 8" (500mm) | 1' 8" (500mm) |
| 4' (1200mm) | 2' 8" (800mm) | 1' (300mm) | 1' 8" (500mm) | 1' 8" (500mm) | 1' 8" (500mm) |
| 4' 8" (1400mm) | 3' 4" (1000mm) | 1' (300mm) | 1' 8" (500mm) | 2' (600mm) | 1' 8" (500mm) |
| 5' 6" (1600mm) | 3' 8" (1100mm) | 1' (300mm) | 1' 8" (500mm) | 2' 4" (700mm) | 2' (600mm) |
| 6' (1800mm) | 4' 4" (1300mm) | 1' (300mm) | 2' (600mm) | 2' 4" (700mm) | 2' 4" (700mm) |
| 6' 8" (2000mm) | 4' 8" (1400mm) | 1' (300mm) | 2' (600mm) | 2' 8" (800mm) | 2' 8" (800mm) |
| 7' 4" (2200mm) | 5' (1500mm) | 1' 6" (450mm) | 2' 4" (700mm) | 2' 8" (800mm) | 2' 8" (800mm) |
| 8' (2400mm) | 5'8" (1700mm) | 1' 6" (450mm) | 2' 4" (700mm) | 3' (900mm) | 3' (900mm) |
| 8' 8" (2600mm) | 6' 4" (1900mm) | 1' 6" (450mm) | 2' 8" (800mm) | 3' 4" (1000mm) | 3' 8" (1100mm) |
| 9' 4" (2800mm) | 7' (2100mm) | 2' (600mm) | 2' 8" (800mm) | 3' 4" (1000mm) | 3' 8" (1100mm) |
| 10' (3000mm) | 7' 8" (2300mm) | 2' (600mm) | 3' (900mm) | 3' 8" (1100mm) | 4' (1200mm) |
| 10' 8" (3400mm) | 8' 8" (2600mm) | 2' (600mm) | 3' 4" (1000mm) | 4' (1200mm) | 4' 4" (1300mm) |
| 11' 4" (3800mm) | 9' (2700mm) | 2' 6" (750mm) | 3' 4" (1000mm) | 4' (1200mm) | 4' 4" (1300mm) |

*Note: 12" (300mm) of soil or a private driveway imposes a load of approximately 5kPa.

## REINFORCEMENT DETAILS WALL TYPE 2

| WALL HEIGHT | | | | REINFORCEMENT | |
|---|---|---|---|---|---|
| TOTAL HEIGHT "H" | HEIGHT OF 150 SERIES BLOCK | HEIGHT OF 200 SERIES BLOCK | HEIGHT OF 300 SERIES BLOCK | X-BARS AND V-BARS | K-BARS |
| 2' 8" (800mm) | 2' 8" (800mm) | — | — | 1/2" (Y12) | — |
| 3' 4" (1000mm) | 3' 4" (1000mm) | — | — | 1/2" (Y12) | — |
| 4' (1200mm) | 4' (1200mm) | — | — | 1/2" (Y12) | — |
| 4' 8" (1400mm) | — | 4' 8" (1400mm) | — | 5/8" (Y16) | — |
| 5' 4" (1600mm) | — | 5' 4" (1600mm) | — | 5/8" (Y16) | — |
| 6' (1800mm) | — | 6' (1800mm) | — | 5/8" (Y16) | — |
| 6' 8" (2000mm) | — | 6' 8" (2000mm) | — | 3/4" (Y20) | — |
| 7' 4" (2200mm) | — | 4' 8" (1400mm) | 2' 8" (800mm) | 5/8" (Y16) | 5/8" (Y16) |
| 8' (2400mm) | — | 5' 4" (1600mm) | 2' 8" (800mm) | 5/8" (Y16) | 5/8" (Y16) |
| 8' 8" (2600mm) | — | 5' 4" (1600mm) | 3' 4" (1000mm) | 3/4" (Y20) | 3/4" (Y20) |
| 9' 4" (2800mm) | — | 6' (1800mm) | 3' 4" (1000mm) | 5/8" (Y16) @ 8" (200mm) | 5/8" (Y16) @ 8" (200mm) |
| 10' (3000mm) | — | 6' 8" (2000mm) | 3' 4" (3000mm) | 5/8" (Y16) @ 8" (200mm) | 5/8" (Y16) @ 8" (200mm) |
| 10' 8" (3400mm) | — | 6' 8" (2000mm) | 4' (1200mm) | 3/4" (Y20) @ 8" (200mm) | 5/8" (Y16) @ 8" (200mm) |
| 11' 4" (3400mm) | — | 6' 8" (2000mm) | 4' 8" (1400mm) | 3/4" (Y20) @ 8" (200mm) | 5/8" (Y16) @ 8" (200mm) |

**ALL BARS ARE TO BE 400 CRS UNLESS OTHERWISE NOTED.**

# Constructing a Beam

Fig. 9.56 shows construction of a beam, either center or external. The external beam is support for a suspended concrete floor. This allows permanent face blockwork to encase the concrete, gives extra support, and does away with an unsightly concrete beam onto which blocks would be laid. It also saves a lot of formwork, which is hard to get true and square at the best of times. Some planning between the bricklayer and the steel fixer is important to ensure steel is placed at the right time and in the right place to engineer's specifications.

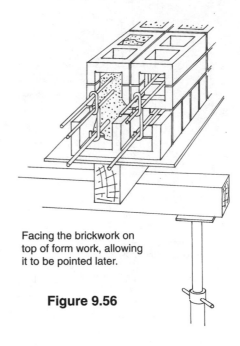

Facing the brickwork on top of form work, allowing it to be pointed later.

**Figure 9.56**

# Rebating to Allow Wall Drainage

Clay blocks used on external weather walls as in the case of a home garage can take in moisture. Weep holes are required to allow the moisture to drain below floor level to the outside. To construct a cavity rebate, nail a piece of timber to the bottom of the block on the floor side below floor level that sits on the DPC, and then wrap the DPC down the inside of the timber and across underneath. The timber is removed after the floor is poured leaving a cavity between the block and floor. Weep holes are pushed through the bottom of the perpendicular mortar joint with a small toothing filler ³⁄₁₆" (6 mm) on the top of the DPC. This allows moisture and water to escape from inside the wall.

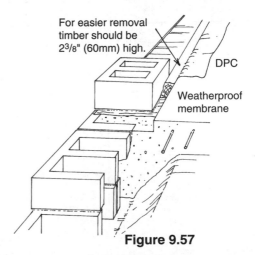

For easier removal timber should be 2³⁄₈" (60mm) high.

DPC

Weatherproof membrane

**Figure 9.57**

# Notes

# Notes

# Part II

# Brickwork — Advanced Construction

# 10

# Introduction to Advanced Construction

## Lintel Construction

### The Four Best Methods of Brick Lintel Construction

Lintel beams were described in the earlier section on brickwork, and they are used to hold up the brickwork courses above doorways and window heads or other openings.

**Method A** — The angle-iron lintel protrudes ½ a brick into the wall at either end of the opening and is exposed. The brick that supports the lintel is rebated, the lintel is placed directly on top, and the soldier course is laid right to the end of the rebate.

**Figure 10.1  A**

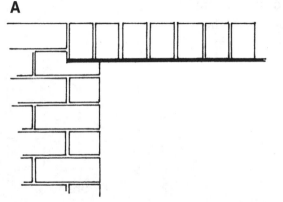

**Figure 10.2**

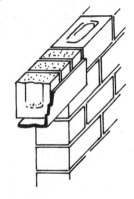

**Method B** — Internal Corner — The brick that bears the lintel is rebated.

In this case the soldiers are laid directly to the back of the internal corner and cut to height to work brick course.

**Figure 10.3**

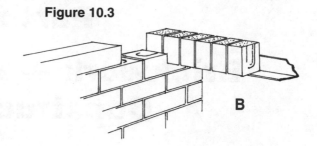

B

**Method C and D** — A more practical method where the face of the lintel is hidden. The brick underneath the lintel is checked out and a brick piece or "biscuit" is laid on top of the lintel. Soldiers are then laid in the opening, butting up to each side of the jamb.

**Figure 10.4**

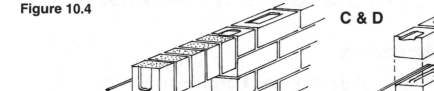

C & D

**Method E** — The lintel bar or arch bar is used in the event of difficult head height—the face of the lintel is hidden. A brick is cut over the top of the arch bar and the brick "biscuit" placed underneath the lintel to stabilize it. Soldiers are then laid directly in the opening butting up to each side of the jamb.

E

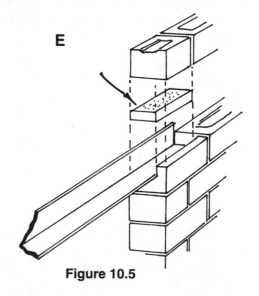

**Figure 10.5**

When laying soldier courses make sure to use an extra line on the bottom of the header to prevent the soldier bricks being laid out of line along the bottom. It is very hard to keep the top and the bottom of the soldier level and plumb at the same time.

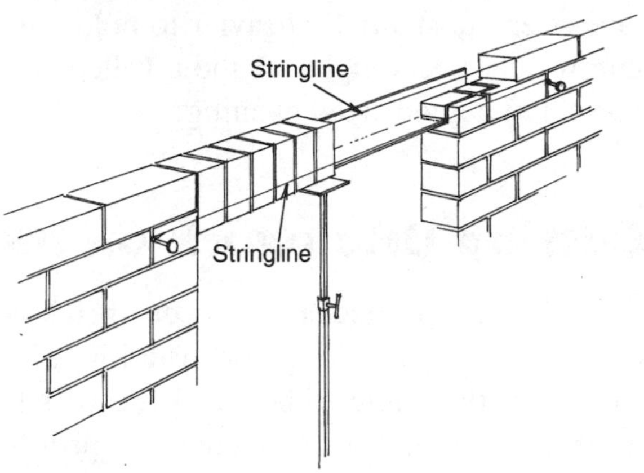

Notice the little biscuits of brick laid on the lintel bars at each end. A little piece of brick is better than mortar because it gives a firm surface to put the next brick on. If we were to fill the rebate with mortar the next brick could sink or tip over. I think it is also worth the extra time to prop the lintel bar.

**Figure 10.6**

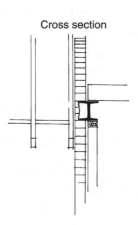

## Arch Supports

**Figure 10.7**

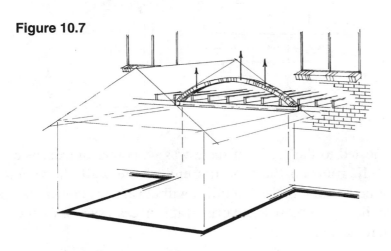

**Figure 10.8**

Cross section

Figure 10.7 illustrates a method I devised to get me out of a spot. It has turned out to be one of the better ideas I have had in terms of cost saving in construction work. The rough gauged arch has been constructed around and onto a rolled steel joist (RSJ) and supports all of the subsequent brickwork up to and around the windows. The RSJ would have had to have been substantially heavier to hold the weight of the wall, but in this way I was able to take the weight off the a rolled steel joist and I didn't need a support post, which left a good sized opening.

# Cutting Brickwork on the Rake

Often it is necessary to cut brickwork on the rake for running up beside steps or in some roofing construction. Figure 10.9 shows two lines. Line A is the line set at the angle to be cut. Line B is the stringline at which the brick courses are to be laid, and is connected to a profile at one end, joining a toothed corner at the other end.

**Figure 10.9**

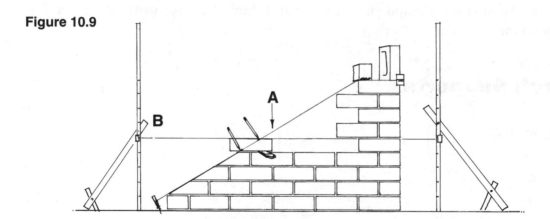

Line A could be connected to the profile if there was no corner in existence. The profile shown in the diagram on the toothing end of the wall shows an alternative method of construction. Using these profiles will always mean there is a stringline to be followed up the brick courses, and even if it is necessary to lay cut or half bricks, the courses will remain consistent.

The brick to be laid to the rake is held in place using a carpenter's rule and the brick is then marked corresponding to the stringline. If they are pressed bricks or frog bricks, the angle cut can be made with a bolster and hammer.

Extruded bricks with holes should be cut with a power saw using a serviceable diamond-tooth blade or masonry wheel.

**Buttresses** – Fig. 10.10, Completely built out of English bond, when the width is to be reduced we can create a gun by a stringline attached to the profiles as in the diagram. The buttress is reduced one brick at a time. Check to see if the tumbling-in works gauge on the incline and courses matching the vertical gauge. The bottom tumbling-in shows the inclined bricks cut into the horizontal courses every third course Fig. 10.11. The top incline where the internal ends of the incline are finished by toothing, is more difficult requiring much more cutting—a time-consuming exercise.

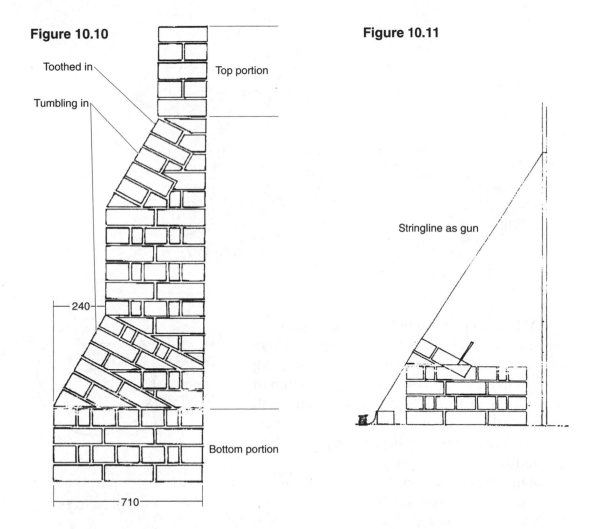

**Figure 10.10**

Toothed in

Tumbling in

Top portion

240

Bottom portion

710

**Figure 10.11**

Stringline as gun

# Brick Ramp and Adjoining Steps

First lay the perimeter of the ramp and steps making sure to bond them in. When constructing the raking walls on the ramp make sure you bond in the brick risers. I find it easier to set up a couple of profiles, to cut the rake of the ramp as shown in Fig. 10.12, and lay the floor of the ramp as shown in Fig. 10.13.

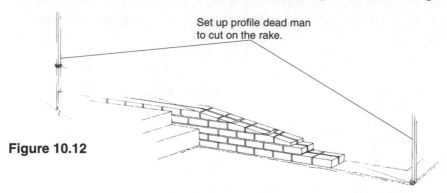

Set up profile dead man to cut on the rake.

**Figure 10.12**

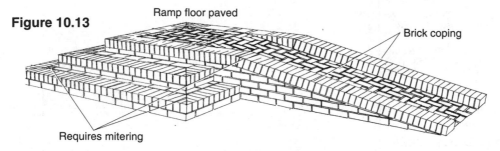

**Figure 10.13**

Ramp floor paved

Brick coping

Requires mitering

**Figure 10.14**

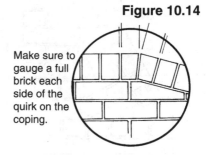

Make sure to gauge a full brick each side of the quirk on the coping.

When you are setting up a job try to use two full bricks at the top of the coping where the ramp ends and the top of the coping platform begins. It is better to cut the bottom of the bricks than to use wedge bricks. This will also make the miter look best. Look also where the level floor meets the angled ramp as well as the inside of the coping. Notice that the corners of the step treads on three sides will have to be mitered. The floor of the ramp is made of solid bricks.

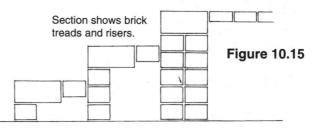

Section shows brick treads and risers.

**Figure 10.15**

# Brick Miter Box

Figure 10.16 shows my invention for cutting brick miters. It is very useful for all external miters and is made quite simply from a piece of checker plate aluminum with a steel square packed up ⅜" (10 mm) and bolted to the checker plate. The reason for the ⅜" rise is to compensate for the rough corners on many bricks which would otherwise not sit straight.

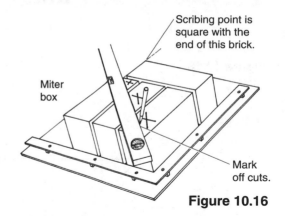

**Figure 10.16**

Place the bricks in the box as shown. Square off brick number one and scribe a line diagonally. Mark the "off cuts" with crosses and repeat for the matching side. This completes the miter. This box allows all miters to be precut by the bricklaying assistant, saving time and money.

# Miter Corners

Miter corners are mostly used on returned corners and as caps on brick piers. Due to the use of extruded bricks, mitering is an ideal method for hiding the holes on the capping course. It also gives a neat and tidy finish to brick pier caps.

Practice makes perfect, and with time your ability to complete miter corners will become well refined. There is little description I can give on how best to go about their construction, except to refer you to the illustrations for single skin, double skin, and brick cap miter cornering. The drawings are self-explanatory and the measurements will guide you through the learning process. All the cutting work in this type of brickwork must be carried out with a diamond saw.

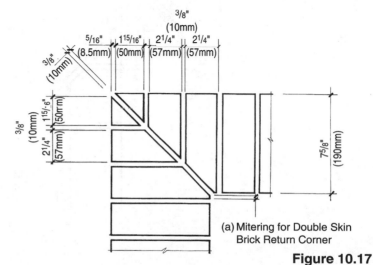

(a) Mitering for Double Skin Brick Return Corner

**Figure 10.17**

**Figure 10.18**

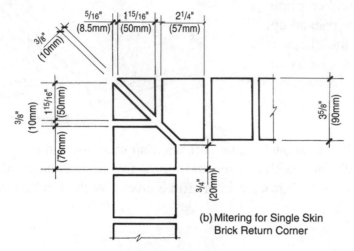

(b) Mitering for Single Skin
Brick Return Corner

For easy cutting set the diamond saw guide
at 45° and it will cut these automatically.

**Figure 10.19**

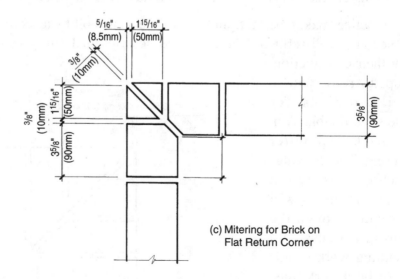

(c) Mitering for Brick on
Flat Return Corner

# Brick Cap for 11¼" (350 mm) Piers

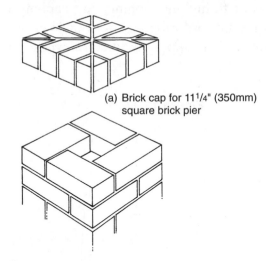

(a) Brick cap for 11¼" (350mm) square brick pier

**Figure 10.20**

Take care when mitering brick caps on piers as it pays to tighten the perp ends up on the pier at the construction stage as in Fig. 10.20a. Be on the look-out for varying brick lengths as this can significantly vary the brick gauge. On a brick pier, say 11⅝" (295 mm) wide, four bricks less one joint is only 11⁷⁄₁₆" (290 mm) so step the cap in ³⁄₁₆" (5 mm) around the top. This fixes the gauge problem and if we round iron joint the cap, all will look well finished.

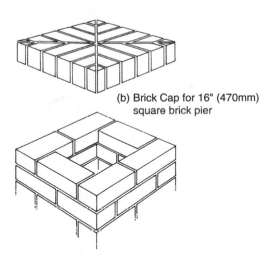

(b) Brick Cap for 16" (470mm) square brick pier

**Figure 10.21**

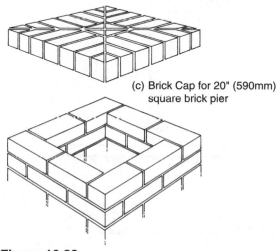

(c) Brick Cap for 20" (590mm) square brick pier

**Figure 10.22**

# Fitting Ant Capping

Figures 10.23 to 10.25 show examples of fitting ant capping. Ant capping is designed as a slow-up to ants (termites) in the upward progress from their ground dwelling to their potential tasty lunch of timber. Ant capping is fitted to expose the ants and acts as a weather deterrent.

**Figure 10.23**

**Figure 10.24**

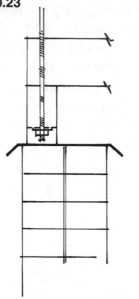

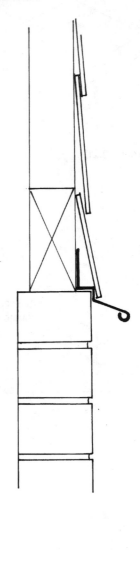

**Figure 10.25**

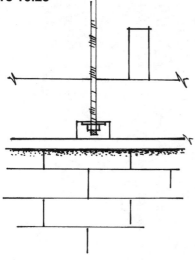

**Figure 10.26**

If you are to fit ant capping, take the time to position it well so it sets off the finished job. It may mean checking out a bearer to house a tie-down bolt that can no longer fit in the cavity, but it also means the nut can be tightened later.

Take a closure look at Fig. 10.26 left and notice how the tie-down rod is fitted in the cavity and anchors the bearer by a homemade bracket.

The use of the simple bracket can save a lot of cutting of bricks that would have had to be done to enclose the rod.

**Figure 10.27**

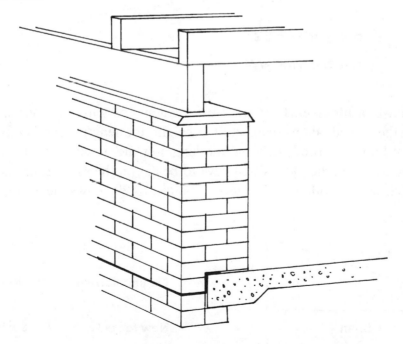

# Working Out a Brick, Block, or Clayblock Gable without Trusses in Place

**Figure 10.28**

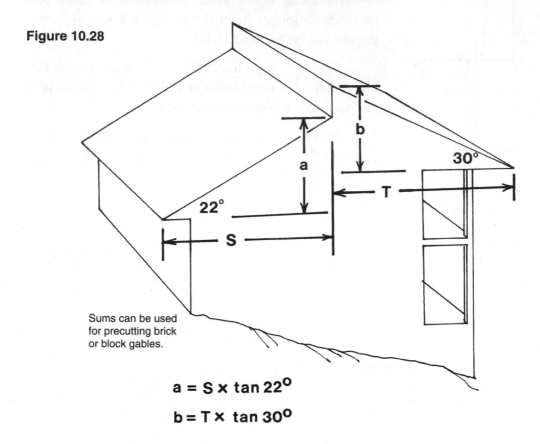

Sums can be used for precutting brick or block gables.

$$a = S \times \tan 22°$$

$$b = T \times \tan 30°$$

This method is a lifesaver. First we need a calculator with sin, tan, and cos functions. Using these methods we can find the height, the length, and the angle of the trusses. If you know the roof pitch and the length or the height of the truss you can find the angle at which the blocks will have to be cut on the rake. There are five calculations, which will provide you the answer. Figure 10.29 shows the examples.

**Figure 10.29**

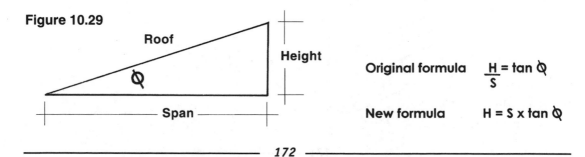

Roof

Height

$\emptyset$

Span

Original formula $\quad \dfrac{H}{S} = \tan \emptyset$

New formula $\quad H = S \times \tan \emptyset$

# Setting Up for Uneven or Sloping Sites Using a Calculator

There is not a lot of difficulty in setting out job sites on sloping ground and you won't need a degree in geometry if you follow some basic rules. It is not always possible to erect profiles because of limited room between buildings and where machinery has to move around to dig foundations.

**Figure 10.30**

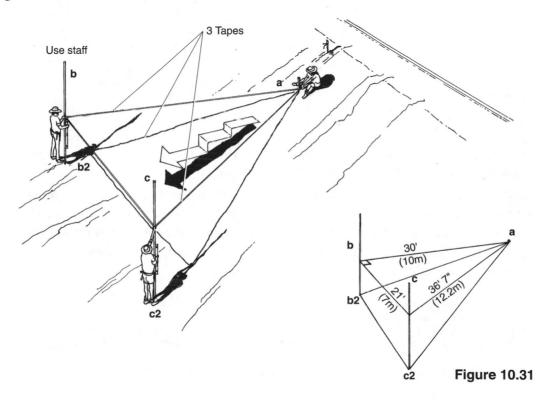

**Figure 10.31**

***Step 1*** Drive a peg off the kerb as a reference. Then position another datum peg (a) about a meter in from the first peg. If for example we were working on a house with dimensions of say, 30' (10 m) by 21' (7 m), we would measure down the site 30' from our datum peg at ground height. Using a second tape measure, we measure across 21'. With a third tape we measure back to our datum peg and the distance must be the hypotenuse, which is the square root of the sum of the squres of the other two sides.

**Figure 10.32**

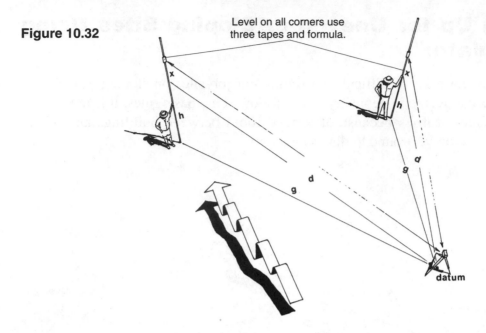

Level on all corners use three tapes and formula.

datum

***Step 2*** So for a house 30' × 21', the third side of our triangle, the hypotenuse, will be 36' 7". Using the hypotenuse ensures a true right angle. Next, using our laser level from our datum point, we shoot a level to our first point (b) and our second point (c). If the level from (a) across to (b) and (c) is three foot then using our hypotenuse mathematics we can determine the length from the ground at (b) to (a). If, as we said, the building is 30' and our level is 3', then the difference from (b2) to (a) will be 30.149'.

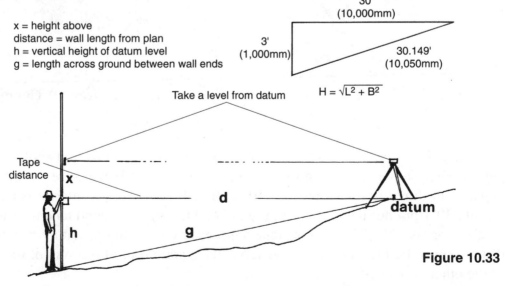

x = height above
distance = wall length from plan
h = vertical height of datum level
g = length across ground between wall ends

30'
(10,000mm)

3'
(1,000mm)

30.149'
(10,050mm)

$$H = \sqrt{L^2 + B^2}$$

Take a level from datum

Tape distance

x

d

h     g

datum

**Figure 10.33**

***Step 3***    Repeat this procedure for the other side, then you will know for sure the distances between (a) and (b2) and (a) and (c2) "on the ground." Now we can put in pegs at (b2) and (c2) and by measuring across our 21' to complete the rectangle we can put in our last corner peg.

***Step 4***    This area is then string-lined and marked out between the pegs with lime for excavation to the required depth of the footing.

**Figure 10.34**

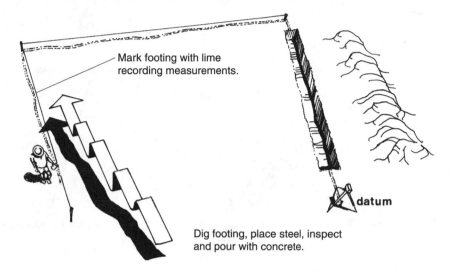

Mark footing with lime recording measurements.

datum

Dig footing, place steel, inspect and pour with concrete.

***Step 5***    Note that you will have to step the footing to work modules of brick or block otherwise you will have a sloping footing. Using our three tapes again and our measurements from before, we mark out our footings for 30' by 21' required.

**Figure 10.35**

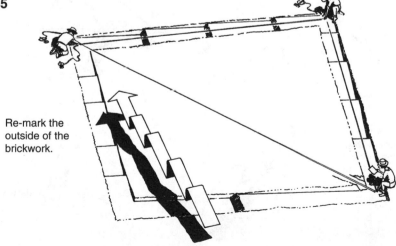

Re-mark the outside of the brickwork.

**Step 6**   Our next step is to set up our profiles on the corner marks. Refer back to our datum point which is ground level at (a), mark the gauge down the profile. This way we can block or brick up our corners and we know they will meet up when we lay in between.

**Figure 10.36**

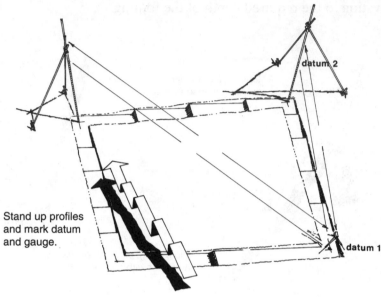

Stand up profiles
and mark datum
and gauge.

datum 2

datum 1

**Figure 10.37**

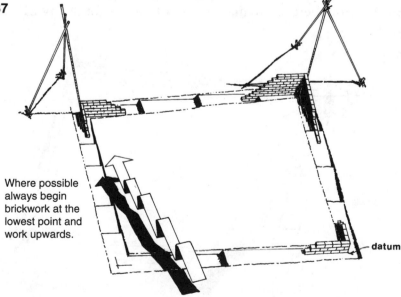

Where possible
always begin
brickwork at the
lowest point and
work upwards.

datum

# 11

# Arches

*Care on the job site is vitally important. Cutoff saws are especially dangerous to the inexperienced. Always cut only with the blade at full revs—it will reduce the chances of "grabbing," If you have to hang on to the brick hold on to the side of it so it will not pull your hands through the blade.*

*I like to see ear and eye protection but do not advise using gloves when cutting bricks.*

## Setting Up

When building arches I prefer to lay the two walls which run at 90 degrees to the arches, resulting in two corners already constructed.

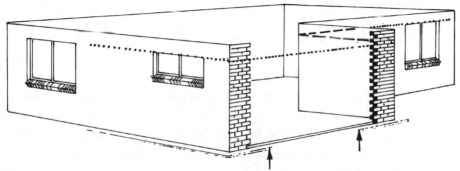

**Figure 11.1**                                    Arrows indicate top of footing.

If the arch centers are already made, place a brick on edge on the centers and measure the distance from the bottom of the centers to the top of the brick. Allow headers each side of the template when measuring.

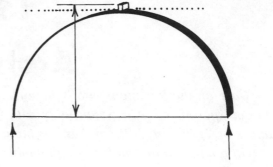

**Figure 11.2**

## Finding the Springing Point or Start of the Arch

1.  Work out how many courses are needed to go around the arch. These are known as the voussoirs. The underside of the header or voussoir bricks should finish near the window or door head height (refer to dotted lines in previous diagrams).

2.  After establishing the overall height of the arch, run a line through the course directly over the voussoir and measure down the distance shown. This is the springing line.

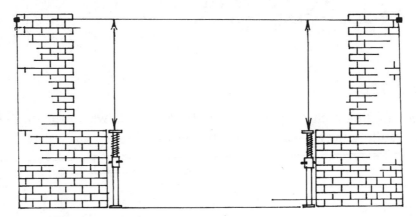

**Figure 11.3**

3.  After you have laid the brickwork up to springing line, measure down from the springing line to the footing. There are two ways of supporting the arch. One is to cut blocks of timber, but this has become outdated with the use of screw-up Acrow props which allow adjustment of either end of the arch, up or down as required.

4. Cut or stand props to the appropriate length. (Allowing for wedges if using timber stays so they can be easily removed.)

5. Cut a spreader from some 3" × 2" (75 × 50 mm) timber for the props, to stop them moving sideways as you work.

6. Place the arch center on the props and adjust to the appropriate height. Sight across the top of the wall in which the arch is to be built to identify any line sag. Moving the arch center up as required will avoid a thick bed joint on the headers at the top of the arch.

**Figure 11.4**

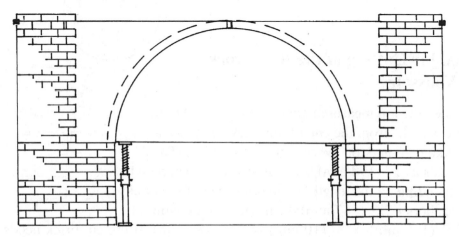

Make sure the arch center is level across the springing line beneath the arch; otherwise the arch could be higher on one side and not a proper semi-circle. Place a brick across the top of the arch center and move the center so the brick is level on top. If you have to, disregard plumbing the arch center to make sure the brick on top is level.

**Figure 11.5**

Run a tape over the circumference of the arch and refer to the scale in this book to find the gauge around the template.

Remember that, laying around a circle, you will have to choose a smaller gauge; otherwise the top of the joint, the perpend between the voussoirs, will be large in respect of the size of the arch.

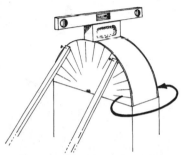

Gauge marks are on opposite side as per arrow.

If the gauge doesn't "work" all the way around the arch, find the center and plumb a line to the top of the arch. Allow half a joint each side of the center line of the arch. Start your gauge on either side of this center line.

A carpenter's pencil laid along the curve of the arch—at the ends where the wall courses adjoin — will assist in determining the length and the angle of the brick to be cut in on stretcher bond to meet the arch.

**Figure 11.6**

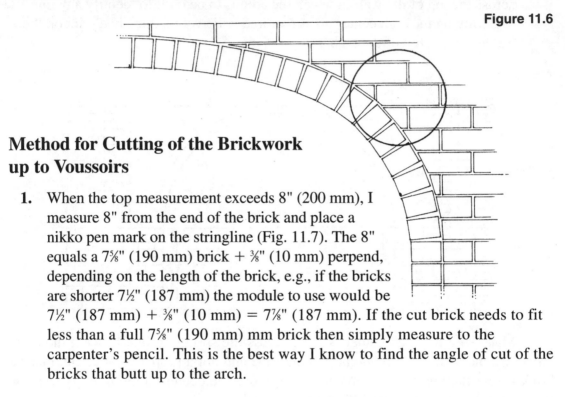

## Method for Cutting of the Brickwork up to Voussoirs

1.  When the top measurement exceeds 8" (200 mm), I measure 8" from the end of the brick and place a nikko pen mark on the stringline (Fig. 11.7). The 8" equals a 7⅝" (190 mm) brick + ⅜" (10 mm) perpend, depending on the length of the brick, e.g., if the bricks are shorter 7½" (187 mm) the module to use would be 7½" (187 mm) + ⅜" (10 mm) = 7⅞" (187 mm). If the cut brick needs to fit less than a full 7⅝" (190 mm) mm brick then simply measure to the carpenter's pencil. This is the best way I know to find the angle of cut of the bricks that butt up to the arch.

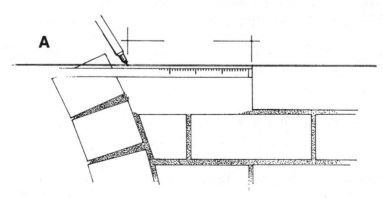

A

**Figure 11.7**

**Figure 11.8**

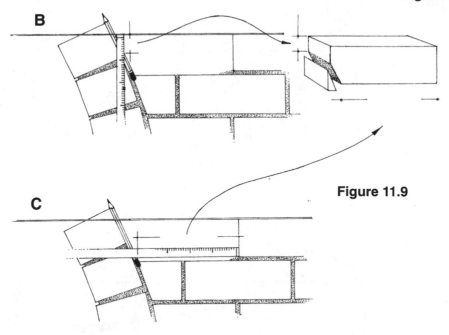

**Figure 11.9**

Semi Circular Arches with Milder Angle Cuts

2.  Place your pencil on the voussoirs as shown, and measure horizontally the bottom of the brick.

    Keeping the ruler straight is most important. Also, make sure you have your ruler up enough to cover the thickness of the mortar bed you intend to lay. This ⅜" is not great, but extended across the angle it can made a lot of difference over the distance of the bed joint. Now deduct ⅜" from the bottom measurement to allow for the perpend on the brick (Fig. 11.9).

3.  Mark the bottom measurement on the brick to be cut. On this brick, measure vertically down from the 8" (200 mm) previously marked on the line to the pencil. This measurement can be transferred onto the brick to be cut. There is no need to deduct ⅜" from the vertical measurements as the pencil compensates for the bed joint around the arch (Fig. 11.8).

4.  A line intersecting both marks can be drawn with a Nikko pen and cut with a saw.

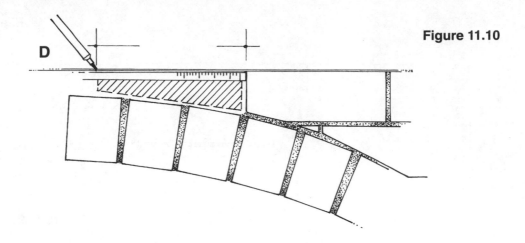

**Figure 11.10**

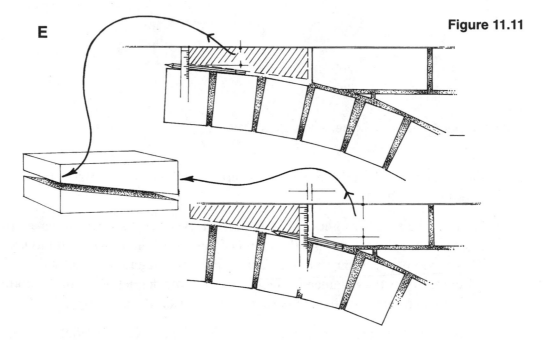

**Figure 11.11**

Flatter Arch — longer angle cuts

5. Measure 8" (200 mm) along the line from the end of the brick. Place Nikko pen mark on the line.

6. Place carpenter's pencil directly under the mark on the line and measure from the line to the pencil. This is the front measurement required, so mark it on the brick to be cut (Fig. 11.11).

7. Place the carpenter's pencil directly in front of the previous brick laid. Measure from the stringline to the pencil but don't allow for a bed because this is taken up by the pencil thickness. This is the rear measurement required. Transfer this length to the brick to be cut and proceed to cut the brick (Fig. 11.11).

## Method for Laying Rough-Faced Bricks Around the Circumference of an Arch

Because of the large variation in the face of bricks, the arch center should have a piece of masonite tacked around it. Lay a bed of lime-only mortar around the arch center on this masonite.

We use a lime-only mortar as a basing in bed for the brick faces so it can be scraped out later and the face cleaned with a scrubbing brush. The mortar between these voussoirs is pushed through from either side with a toothing filler tool to give the familiar raked joint effect.

The voussoirs can be laid around the arch center disregarding the rock face. Be sure the back edge of the voussoirs is on an even curve by measuring off the arch center to the back edge of the bricks.

**F**

**Figure 11.12**

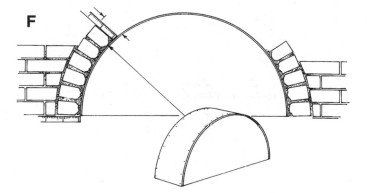

# Arch forms

## A More Rounded Arch

**Figure 11.13**

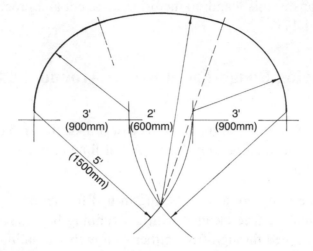

By maintaining the diameter and changing the radii of a semi-elliptical arch, a more rounded arch is formed (Fig. 11.13). This enables the voussoirs to be laid around the center without the wedge joint effect often encountered on the tight radius of the semi-elliptical arch.

## Semi-Elliptical Arches

**Figure 11.14**

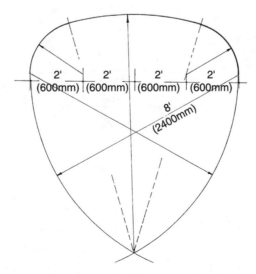

**Figure 11.15**

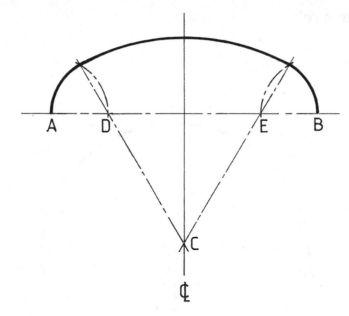

Simple method for scribing arch center on building site

1. Draw springing line AB and bisect

2. Mark rise CD

3. Using C as center and AD as radius, intersect springing line at points E and F

4. Pin string at points E and F

5. Length of string equals AB

6. Place pencil on inside of string and scribe intrados

By moving points E and F different forms can be obtained

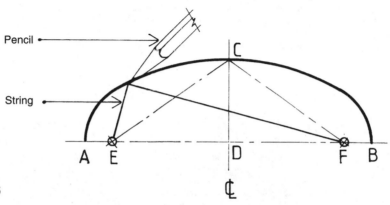

**Figure 11.16**

## Semi-Circular Arch

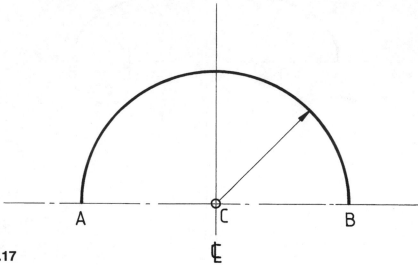

**Figure 11.17**

1. Draw a line and mark off the span of the arch (AB)
2. Find the center of the span, C
3. Using point C as center, scribe the two arcs AB

# Pointed Trefoil Arch

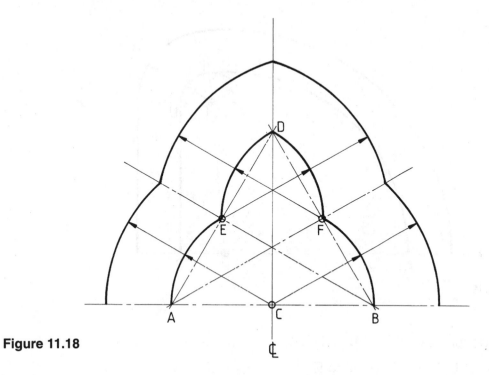

**Figure 11.18**

1. Draw springing line AB and bisect

2. Draw 60° ▲ADB

3. Draw lines AF and BE 30° above line AB

4. Using C as center, scribe arcs AE and BF

5. Using E and F as centers, scribe arcs DE and DF

6. Mark arch face 7⅜" (190 mm) and repeat 4 and 5

7. Set out arch bricks on extrados and draw in gauged and bonded face

# Rampant Arch

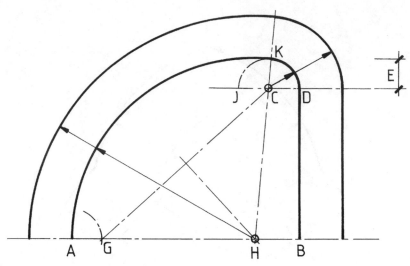

**Figure 11.19**

1. Draw springing lines AB and CD

2. Cut off DC and AG equal to E

3. With C as center and E as radius, scribe semi-circle DJ

4. Join points CG and bisect, produce this line to intersect AB at H

5. Draw common normal HK through H and C. With H as center, radius HA, scribe intrados AK

# Arabian Arch

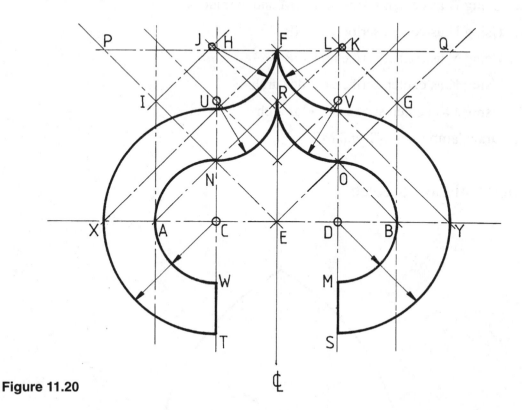

**Figure 11.20**

1. Draw springing line AB and bisect
2. Draw 45° ▲ARB
3. Bisect lines AE and EB
4. Mark *X* and *Y* 7⅜" (190 mm) out from A and B, respectively
5. Draw 45° ▲XFY
6. Draw two 45° lines from E (1 each side of CL)
7. Draw a horizontal line through F to cross lines drawn in 6
8. Bisect lines RO and RN and produce to line PQ
9. Draw vertical lines through points A and B
10. Draw 45° lines through points G and I
11. Draw 45° lines through points L and H to cross lines drawn in 10

12.  Using C as center, scribe arch WN and extrados arch

13.  Using D as center, scribe arch OM and extrados arch

14.  Using U as center, scribe arch NR

15.  Using V as center, scribe arch OR

16.  Using K as center, scribe extrados arch

17.  Using J as center, scribe extrados arch

18.  Draw jamb lines WT and MS

# Equilateral Gothic Arch

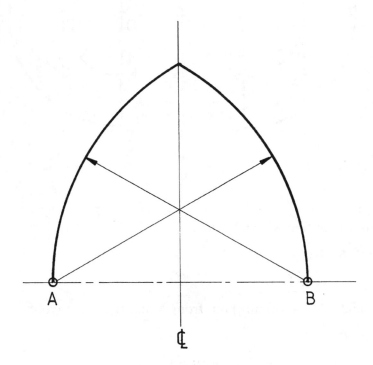

**Figure 11.21**

1.  Draw springing line AB 20" (500 mm) and bisect

2.  Using centers A and B scribe arcs to CL

3.  Mark arch face 7⅜" (190 mm) and repeat from centers A and B

4.  Set out arch bricks on extrados with a multiple of 4 + 1

5.  Using centers A and B draw in arch face

# Tudor Arch

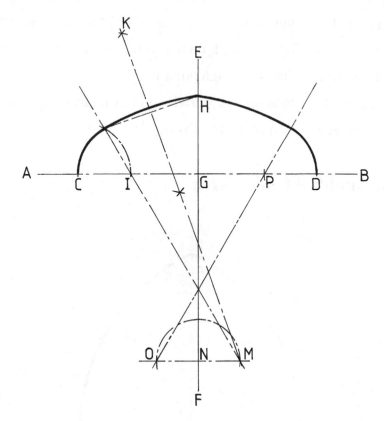

**Figure 11.22**

1. Draw the springing line AB

2. Mark off the span on this line, points C and D

3. Bisect the span with line EF, crossing line CD at point G

4. Mark off the rise on line EF, from G to H

5. Mark off two-thirds of the actual rise, from point C, along the springing line to point I

6. Using point C as center, scribe an arc from point I to J. Using point I as center, scribe an arc from point C to J

7. Draw a line from point J to point H

8. Bisect line JH, carrying it well down to point L (line KL)

9. Draw a line through points J and I, crossing line KL at point M

10. Draw a line, parallel to the springing line, from point M, crossing line EF at N. Carry this line past the center line, about equal distance to MN

11. Using point N as center, scribe an arc from point M to cross line MN at point O

12. Mark off an equal distance to CI, on the springing line, from point D to P

13. Draw a line from point O, through point P

14. Use centers I and P to scribe the intrados and extrados for the hunches

15. Use centers M and O to scribe the crown of the arch

## Dropped or Modified Gothic Arch

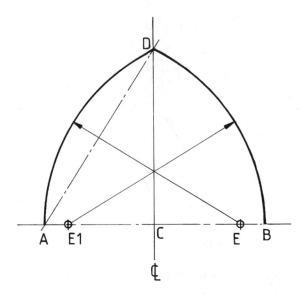

**Figure 11.23**

1. Draw springing line AB 2' (600 mm) and bisect

2. Mark rise on center line CD 16" (400 mm)

3. Bisect AB and produce to springing line for striking point E. Make E2 equal to CE

4. Using E and E2 as centers, scribe arcs to center line point D

5. Mark arch face 7⅝" (190 mm) and repeat from centers E and E2

6. Set out arch bricks on extrados and draw in gauged and bonded face

*Note: To maintain a stretcher at the springing point and crown, the multiple of bricks must be 4 + 1.*

# Lancet Gothic Arch

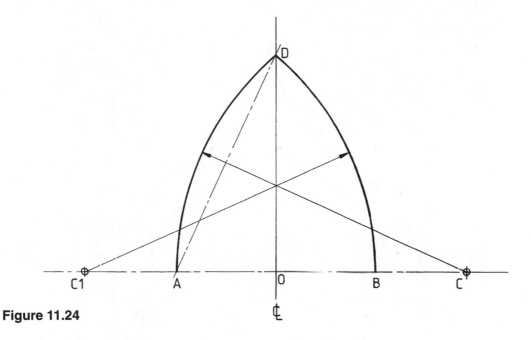

**Figure 11.24**

1. Draw springing line AB 20" (500 mm) and bisect

2. Mark rise on center line (point D)

3. Bisect line AD and produce to springing line for striking point C. Make C1 equal to C0

4. Using C and C1 as centers, scribe arcs to center line point D

5. Mark arch face 7⅝" (190 mm) and repeat from centers C and C1

6. Set out arch bricks on extrados keeping bricks at the crown the same size as others

7. Draw in gauged face from centers C and C1

# Ogee Arch

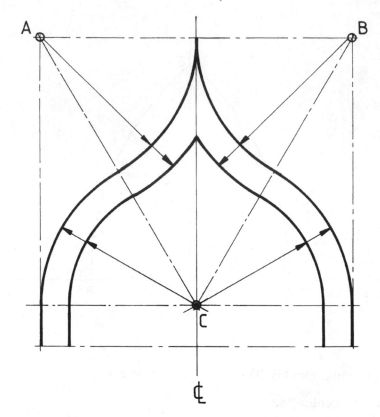

**Figure 11.25**

## Alternate method where no rise is given but span and arch face are known [3⅝" (90 mm)]

*Note: All Ogee Arches are set out on extrados. Height is worked from apex not rise.*

1. Draw square 4' (1200 mm) wide = span 40⅜"+7½"+3⅝" (970+190+90 mm)

2. Bisect square to establish CL

3. Using points A and B as a radius, mark point C on CL

4. Draw common normals through points AC × BC

5. Striking points A, B, and C

6. Decrease radius 3⅝" from point C and draw in intrados

7. Increase radius 3⅝" from points A and B and draw in intrados

8. Springing line is taken from point C

## Dorsal Fin Arch

Can be used as a feature doorway, or as an attraction in bar or entertainment areas.

The arch has to be constructed in a way that is completely viable so a person can walk through it without ducking and weaving. The dot on the diagram shows this point as 6'4" (193 cm).

Following the measurements in the diagram will ensure the arch is structurally sound. It's good practice to lay brick weld into the voussoir and along the matching brick bed joints as much as possible, and to ensure that there is at least three to four courses on top of the arch.

**Figure 11.26**

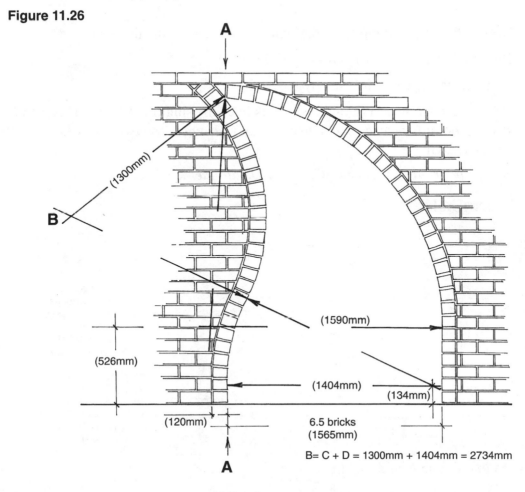

B= C + D = 1300mm + 1404mm = 2734mm

# Segmental Arch

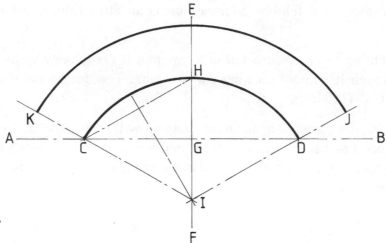

**Figure 11.27**

1.  Draw a line AB and mark off the span of the arch on this line CD

2.  Bisect CD with a line EF, crossing CD at G

3.  Mark off the rise on this line, from G to H

4.  Draw a line from H to C, and bisect this line, crossing line EF at I. This is the striking point for the arch and, using point I as center, the two arcs CD and KJ can be drawn

    *Note: The triangular areas ACK and BDJ are the skewbacks*

# Arch Terminology

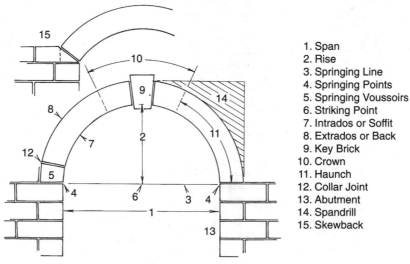

1. Span
2. Rise
3. Springing Line
4. Springing Points
5. Springing Voussoirs
6. Striking Point
7. Intrados or Soffit
8. Extrados or Back
9. Key Brick
10. Crown
11. Haunch
12. Collar Joint
13. Abutment
14. Spandrill
15. Skewback

**Figure 11.28**

# 12
# Bay Windows

## Introduction

To save time I've built my own template for marking out bay windows. It's made out of 2" × 1½" (50 mm × 35 mm) pine and is cleated with ply on the corners to keep it from moving out of square.

The returns on either side of a bay window are always the same angle—45 degrees. My R/H triangular template is pre-made at 23" (575 mm). This allows me to fit a standard 2' (600 mm) window on either side of the window bay.

I stringline across the front face of the wall where the window bay is to be fitted and then plumb down with a spirit level to the concrete footing.

When I have my mark on the footing I lay the level along the ground and sight down over the stringline.

This is preferable to plumbing down in two places. Next I draw a line across the opening.

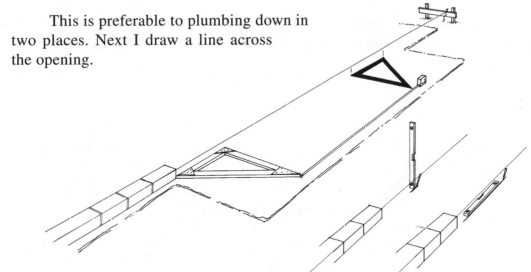

**Figure 12.1**

I always try to make my brickwork "work" up to where the bay window will start because it is untidy to cut bricks before openings and internal or external corners. It makes it look like you didn't plan the job.

I position the template as in Fig. 12.2, with the toe of the template touching the end of the brick run where the internal corner will start. Mark a line from the internal to external corner along the long side of the template. Measure across from the long side of the template the size of the window and add 5⅝"(144 mm). (This works for all window sizes and the attached tables will save you a lot of on-site arithmetic—see formula below Fig. 12.2.)

**Figure 12.2**

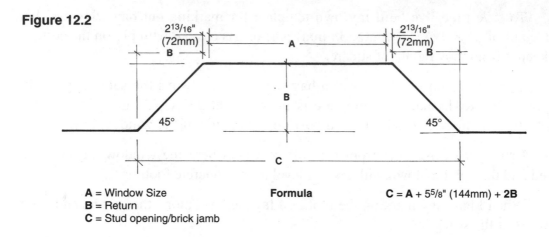

A = Window Size
B = Return
C = Stud opening/brick jamb

**Formula**

$C = A + 5^5/8" (144mm) + 2B$

Relocate the template to the opposite side in reverse and scribe down the long side. The drawings make this a little clearer. The front of the bay can now be bricked up.

**Figure 12.3**

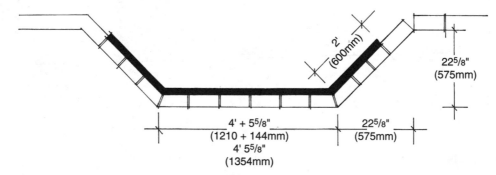

When it comes time to brick up the wings or sides of the bay you will notice the courses won't "work," that is, it will be necessary to cut a brick to fit in. However, this is one small drawback to using standard size windows.

Depending on the bricks, it may even be necessary to cut bricks across the front of the bay.

In these cases I prefer to shorten the bricks on the squint corners to avoid having thick perpends in the face brickwork.

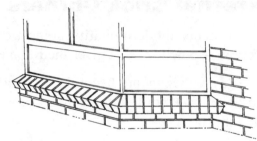

**Figure 12.4**

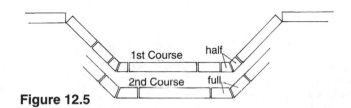

**Figure 12.5**

Variations in brickwork or corners are much less noticeable than bricks cut short to fit under windows.

Figure below shows how a bay window is brought back to fit in behind the outside of the frame. This is achieved using a template with a 21½" (545 mm) measurement as opposed to 22⅝" (575 mm) side.

**Figure 12.6**

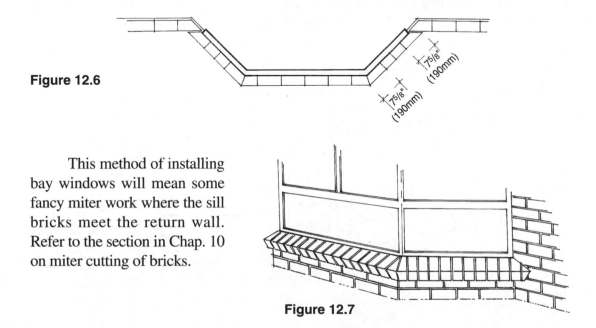

This method of installing bay windows will mean some fancy miter work where the sill bricks meet the return wall. Refer to the section in Chap. 10 on miter cutting of bricks.

**Figure 12.7**

# External Brick Corners on Bay Windows

In many modern quality homes, owners are looking for something a little different—something that gives their home a distinct style or look.

Bay windows are becoming increasingly popular and a nice different touch is to have brickwork on the squint corners of the bay. My method is to work in 2' (600 mm) modules.

First I measure the window and then add the length of two bricks—one for each side of the window. This will be the outside face wall measurement. Add 2' to each side to find the stud opening.

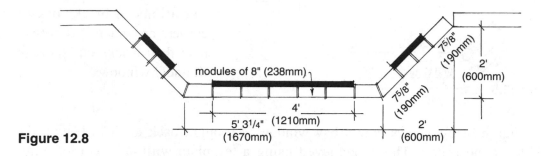

**Figure 12.8**

This additional 2' on each side allows the incorporation of the brick wing walls and the fitting of some custom-made windows. Using squint corners gives the bay window a nice even look and it saves miter cutting bricks along the sills.

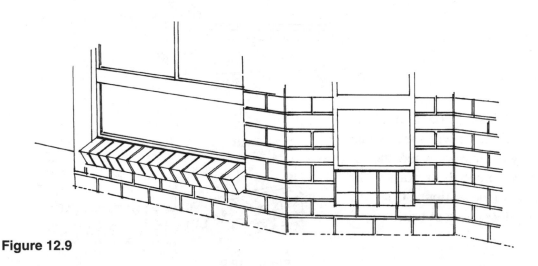

**Figure 12.9**

Fig. 12.10 shows brickwork continuing over the top of a bay window where a small roof is to be constructed. Position the galvanized lintel on the brickwork (supported on each end) then lay a course over the lintel. Cut the building paper and slide the DPC up and under the paper. Let the DPC hang about 3⅛" (80 mm) down over the face of

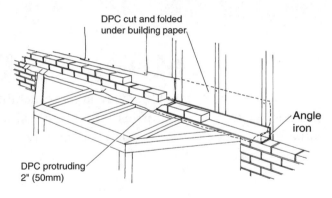

**Figure 12.10**

the brickwork. Weep holes are left on each side of the window up about 8" (200 mm) or to the closet brick pass. The DPC is tacked up inside the roof trusses and the ends trimmed when the roof is finished.

**Figure 12.11**

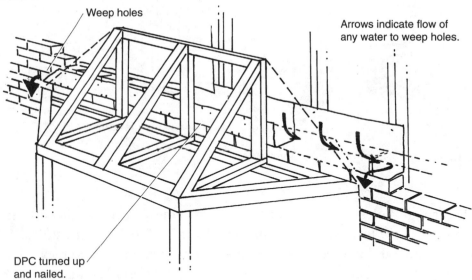

# 13

# Setting Up Profiles

This is the key to efficiency and profitability and external profiles are perhaps the easiest to set up. Profiles which have slides and gauges are the best to use although more expensive to purchase initially.

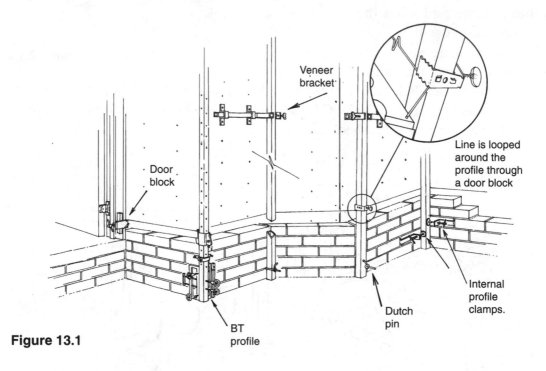

**Figure 13.1**

The gauge holes are prebored and are extremely accurate. It is difficult for some tradespeople to make gauge in modules of ⅟₃₂" (0.05 mm) unless using a precision gauge rod. I have three gauge rods about a meter long and they all have the gauges marked on them. We use them against standard profiles and this makes life much easier. There are, however, some points to remember when using profiles. If the base of brickwork has already been constructed but has not been plumbed true, you may have difficulty setting up your profile correctly, but profiles can be adjusted to compensate for small amounts of hard or lean. When set up, profiles are a temporary corner, for height, plumb, and gauge. It is far easier to run up the brickwork where a profile is assisting you to compensate for hard or lean but nearly impossible to run up using a level especially if the runout is only a few millimeters. Take for example if the brickwork has to be "sneaked in" out of plumb to the inside of the facia as shown. I would say it is nearly impossible to use a level "out of plumb" to arrive behind an object without the help of a straight edge or stringline. The uses for profiles on jobs are seemingly unlimited and they are especially useful when doing inverted corners where the brickwork has to be laid flush with the inside of the facia. I prefer to use the aluminum straight edge thus clamping it to the facia and the brickwork. The end result is that the brickwork ends up behind the facia.

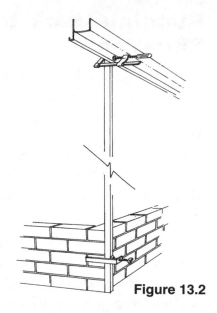

**Figure 13.2**

Notice how the profile is set up in Fig. 13.1. This requires the profile to be set up plumb exactly on the point on the perpend joints otherwise the wall will lean and the perpend joints get smaller as you progress. Proper use of line holders is most important (Fig. 13.4) as is the plumbing of the profile. Fig. 13.3 shows a deadman profile set up in a doorway.

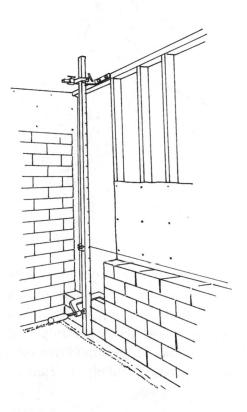

**Figure 13.3**

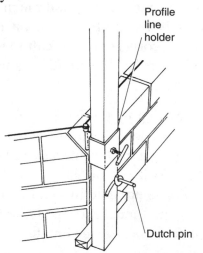

Profile line holder

Dutch pin

**Figure 13.4**

# Running Back to a Door Frame in a Veneer Situation

First mark up the gauge on the door frame then attach a "door block" as shown in Fig. 13.1. Door blocks are available from most hardware suppliers and make life simple for bricklaying up to doorways where it's difficult to follow a stringline. They save a lot of time in checking that courses are level. Taking a closer look at the block we can see it has been designed to allow the brick to be laid under the block without shifting the line up. These same blocks are modified and can also be used on the profile posts or on the steel door frames that are made for single skin brickwork or block work.

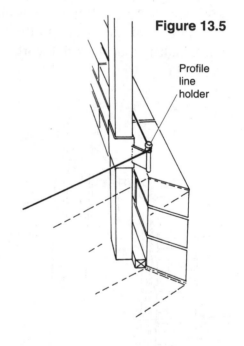

**Figure 13.5**

Profile
line
holder

Any leaning either in or out will tighten or thicken the squint joint. I designed these slides to fit over the standard aluminum profiles for especially difficult corners. They can be used on veneer construction and with the help of a few props take away a lot of headaches in a brick base construction where the brickwork has to be set out on the footing as in Fig. 13.5.

It has always been my preference to lay blocks from the outside. I find it far easier to stack the blocks ready for use, they don't have to be carried through the building and it is easier to set up on the outside. External corners are difficult to plumb on the outside, from the inside, especially over waist height.

Setting up profiles is by far easier on the outside and working from the outside means we don't have to wheel the mortar under the scaffold or through doorways. Laying from the outside also ensures the best finish is attained and any imperfections in the blocks on the inside can be easily disguised by round ironed jointing. It is virtually impossible to make a double-sided block or brick, so the bricklayer must give some thought to the desired finish. The inside is the least likely to show up imperfections after the furniture is placed and the curtains are up.

# 14. Suspended Slab

## Using Brickwork for Formwork

Figure 14.1 shows two courses of bricks stepped over 1%6" (40 mm), the distance of the cavity. These two courses can be common bricks usually laid with a bed joint and pushed together with no perp joint allowing the bricks to be removed easily. They can also be painted inside with form oil. This method saves a lot of external formwork, which no matter how hard we try always seems to bow out. I find this method to be a huge saving in both time and money. Having the brick work continuing through the outside of the floor looks much better as well.

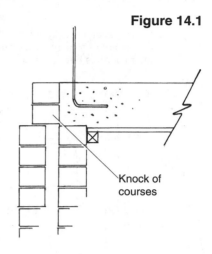

**Figure 14.1**

Knock of courses

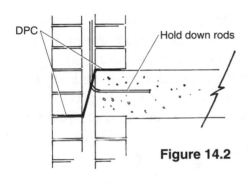

DPC

Hold down rods

**Figure 14.2**

Figure 14.2 shows the fitting of DPC after the brickwork continues. Figures 14.3 and 14.4 show the same method but using brick veneer on the slab. Figure 14.2 shows the DPC and the position of hold-down bolts.

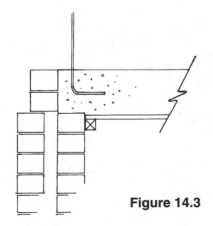

**Figure 14.3**

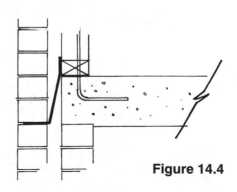

**Figure 14.4**

Figure 14.5 shows how foam can be placed in the cavity prior to pouring the concrete. After the concrete is cured the foam is removed by dissolving with petrol (Figure 14.6). Notice how the DPC can be placed up on one course on the inside skin of brickwork.

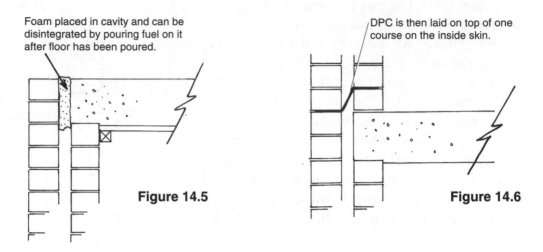

Foam placed in cavity and can be disintegrated by pouring fuel on it after floor has been poured.

**Figure 14.5**

DPC is then laid on top of one course on the inside skin.

**Figure 14.6**

In Figure 14.7, I have shown what I consider to be the best and safest way to load a suspended floor. (1) Place the pallets near the outside load-bearing walls and then load or (2) place the pallets directly on top of the wall underneath. Load up both sides and then scaffold can be erected on the opposite side of wall and loaded as the pallets are placed. Suspended beams as shown in (3) can also be used to support pallets of bricks or blocks. If there are no beams or load-bearing walls, place props underneath the loaded areas (4).

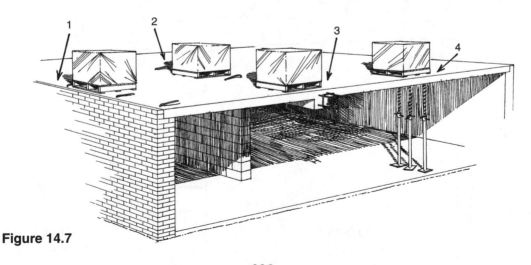

**Figure 14.7**

Remember, there are 66 blocks to the ton and most pallets contain 90 blocks. It's not unusual to see 15 to 20 pallets of blocks loaded on a floor, which means there is up to 33 tons of blocks on the floor. Add to this 6 yards of loam (another 7 tons) plus a ton of cement and the bricklaying equipment such as mixers and scaffold and you can see there is an enormous weight to be placed on the floor. This means you must be sure you plan it well. You should also remember that if it rains and the blocks get wet, they double in weight so cover them carefully as shown in Fig. 14.7.

Always consider the deflection that may take place in a suspended floor during and after a wall has been laid. Loading the floor to introduce deflection without propping may be necessary. Very complex issue and requires careful thought.

To help in building a suspended floor over a creek or a carport I use the following method to get good strength economically. Figure 14.8 shows construction with 8" (200 mm) concrete masonry block work. The voussoirs are 8" × 4" × 4" (200 × 100 × 100 mm) blocks. Spandek is used on a knock out block to allow steel on either side. Spandek is a strong construction material that will be left in place. The retaining blocks are 6" (150 mm) E blocks, which sit on top as shown in Fig. 14.8. Concrete is then poured into the block work and then the deck. The curb blocks are then laid on top as a coping. This method creates a good tie with vertical and horizontal steel.

**Figure 14.8**

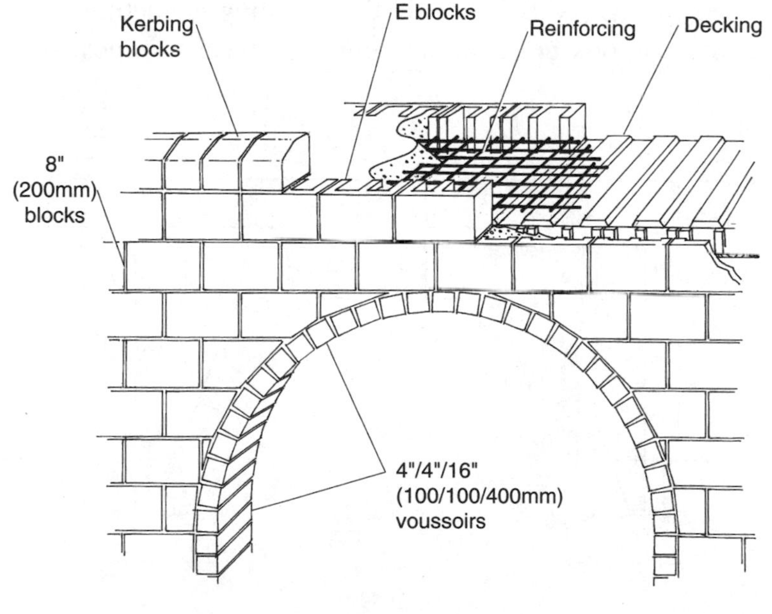

207

# 15

# Fireplace Construction

There is nothing more cosy than a warm fire to sit beside on a cold winter's night. It is, however, hard to relate to this when building fireplaces in the sweltering heat of the tropical summers we encounter in this part of the world.

But fireplaces are becoming more and more popular—to some extent because of their use in winter—but also because of their link with the "old world charm of yesteryear" and the value they add to a home. I can certainly understand their value in homes in cold climates.

The fireplaces in this section are the result of many years in the field, and trial-and-error testing of as many types of fireplaces as possible. The end result was to have a design that had a combination of the most efficient workings, e.g., firewalls, smokeshelf, throat, flue, setout, and simplicity.

## Setting Out

1. Lay three courses around the base as shown by the plan view below. The total floor area of the fireplace is referred to as the recess. In some houses the fireplaces will be recessed out, that is, built external to the outside wall of the construction. Fireplaces are often recessed in, that is, built inside the dwelling, sometimes as a feature and sometimes because carports or adjoining buildings make it impossible to have them recessed out.

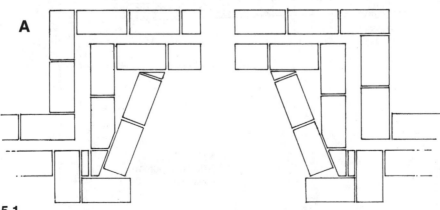

A

**Figure 15.1**

2. Lay the two walls in the fire box. This can be done by fixing a straight edge to the three courses of the base and connecting a stringline to a straight edge or profile post across to connect with a line pin to the three courses previously laid (Fig. 15.2). If you refer back to the segment on "cutting on the rake" you will remember about setting up profile posts and running the stringlines. Adjust the stringline to the correct angle and lay the two walls, cutting to the correct angle.

**Figure 15.2**

3. Place a piece of Hardiplank over the rear of ashpit entrance. The back of the fire box can now be laid (Fig. 15.3).

**Figure 15.3**

I find the use of a "batter-board" (Fig. 15.2) to "run the courses up" when constructing the back of the fire box is less time consuming. It's also more practical because it omits a lot of freehand level work. The gauge can be marked up the center of the board as a guide to keep the bed joints even.

Make the board in two parts to allow easy access of laying. The second board can be slipped in at a convenient height. With two pieces it is also easier to get the boards out again. Lay the rear fire wall, making sure to fill behind it with bricks and mortar and complete the smoke shelf by "dishing it" as shown. Overburned solid bricks are best for these internal walls because they have already been subjected to great heat and seem to cope with the job better.

**Figure 15.4**    **Method of Obtaining Reflection Back Angle**

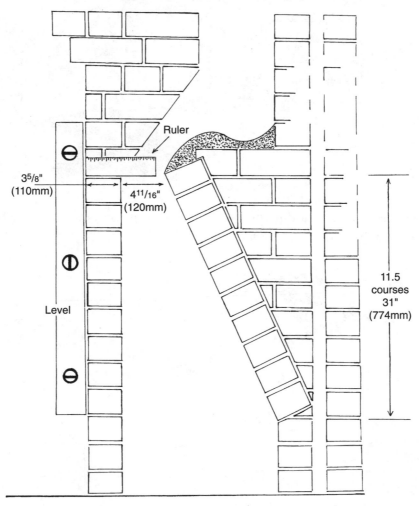

4. If there is no access to the rear of the fireplace, the filling in of the smoke shelf and the rear of the chimney wall must be completed from the front during construction. However if access is available from both sides, try to get a helper to work from the other side towards you — one constructs the firebox, the other fills the rear of the fireplace while progressing.

**Figure 15.5**

5. Next move to the front of the fireplace, and depending on the opening size, lay the brickwork up to the forming of the throat. The use of an arch does away with the use of a lintel bar which could expand and contract with the temperatures and produce cracking.

# Forming the Throat

When forming the throat, the best bricks to use are frog bricks as they have no extrusions and allow better cutting and parging of the throat. (Parging is the plastering of the brickwork inside the fireplace from the smoke shelf up to the chimney stack. This plastering is done with a diluted mortar mix that contains additional lime.)

I find the building of a mantle makes the forming of the throat easier, in so far as it gives more room to bond the brickwork and tie it together.

6. The throat and the gatherings of a fireplace can be constructed to form almost any angle with complete ease if the throat is cut without the use of arch bars, and queen closures are used for the continuation of good bonding.

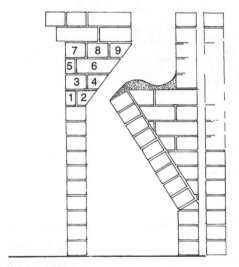

**Figure 15.6**

The correct cutting of bricks is also very important to this process.

The front of the throat can be formed in four courses as illustrated.

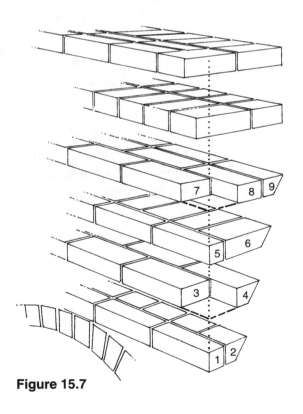

**Figure 15.7**

**Figure 15.8**

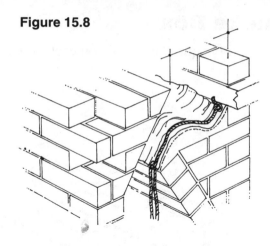

Lay a cleanout rag made of viscreen plastic on the smoke shelf, with a piece of cord connected to each side at the top and bottom.

This rag is used as a temporary measure until the chimney stack has been completed, then removed by pulling the height cords on both sides at the top of the smoke shelf.

This pulls the viscreen with the droppings into the center of the fireplace, leaving a clean smoke shelf.

The angle and measurements are illustrated for fireplace throat.

**Figure 15.9**

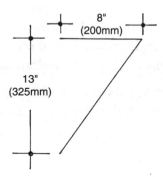

# Smoke Box

Gently square the walls on each side making sure to bond all four walls of the smoke box.

**Figure 15.10**

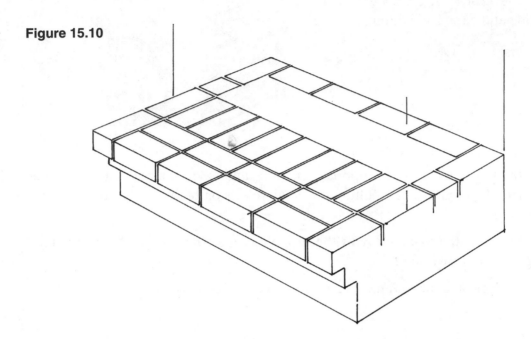

When forming the smoke box, the brickwork has to arrive at the appropriate flue size within a reasonable height, while maintaining adequate flue angles to allow the easy exit of smoke, while preventing a downdraft and the entry of rain.

The flue should also travel its own width from one side to the other and daylight should not be seen through the flue from either above or below.

**Figure 15.11**

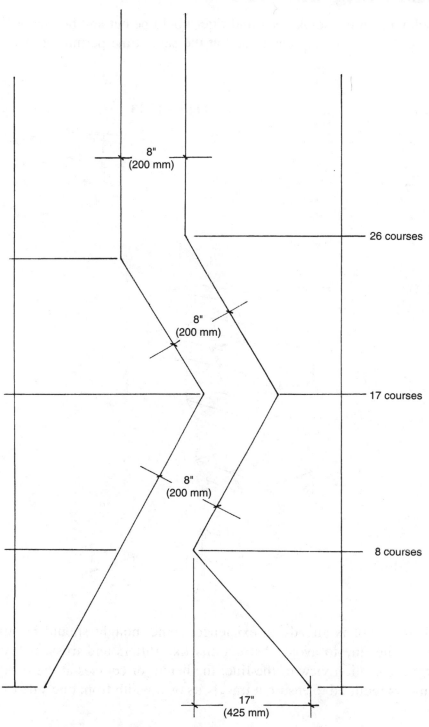

# Constructing the Flue

Brickwork in the smoke box and flue should be cut and bonded as illustrated, keeping the flue structurally sound and at the same time permitting it to be laid to greater heights.

**Figure 15.12**

**Figure 15.13**

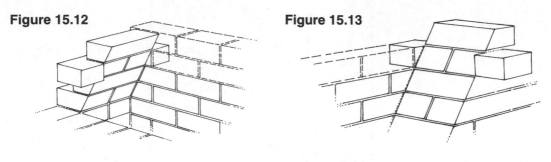

**Figure 15.14**

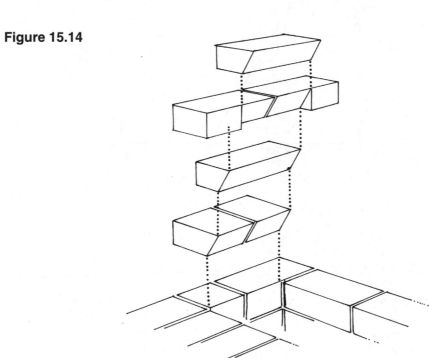

Where a roof is already in existence, some thought should be given to the direction of the flue to avoid obstructions like rafters and trusses. My diagrams show a front elevation view of the flue; the height of courses at the bends; and the measurements required to ensure it travels its own width from one side to the other.

**Figure 15.15**

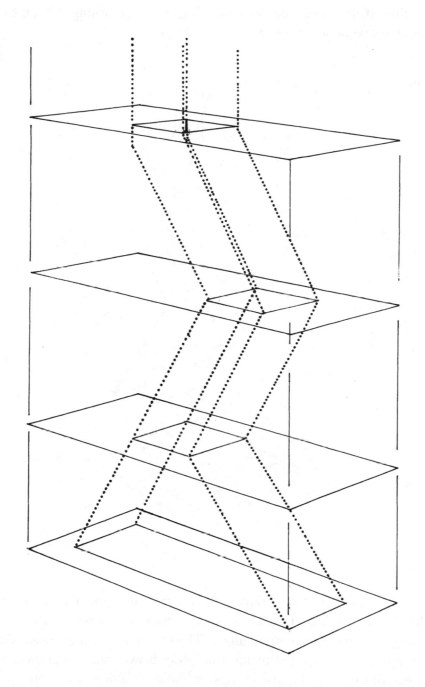

## Cutting the Smoke Box into the Flue Size

I find that, if the smoke box is four bricks wide, returning it back to one brick square should not be done too steeply nor too high.

**Figure 15.16**

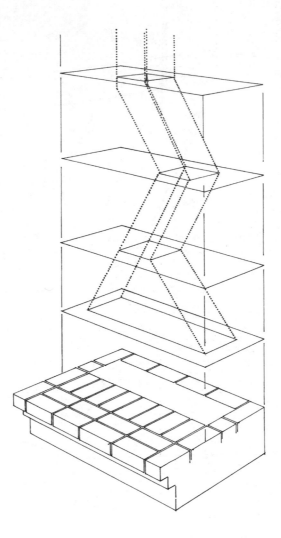

Six courses is enough to arrive at the correct flue size. Continue bonding the bricks around the fire box. Use king closers, making sure to cut these and their corresponding headers at the correct angle. The internals and the smoke shelf, smoke box, throat, and flue are parged using a pointing trowel and bagged off with a wet sponge to smooth off any lumps, bumps, or holes, leaving a smooth glassy finish inside the flue.

**Figure 15.17**

# Position of the Fireplace in a Dwelling

Stories abound about problems with fireplaces. Usually any problems with a fireplace can be overcome by sensible planning at the outset. Location of the fireplace in a room is important.

Moreover, location of windows, doors, and openings in relation to the position of the fireplace are important considerations when planning a house that will have a fireplace. Air flowing through windows as seen in Fig. 15.18 can cause a fireplace to smoke.

The solution is to close the openings.

However, some people like a free flow of air through the room with the fireplace.

Construction of the room as in Fig. 15.19 allows fresh air to enter the room without affecting the fireplace.

**Figure 15.18**

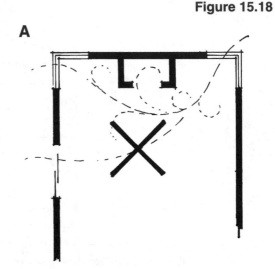

**A**

**B**

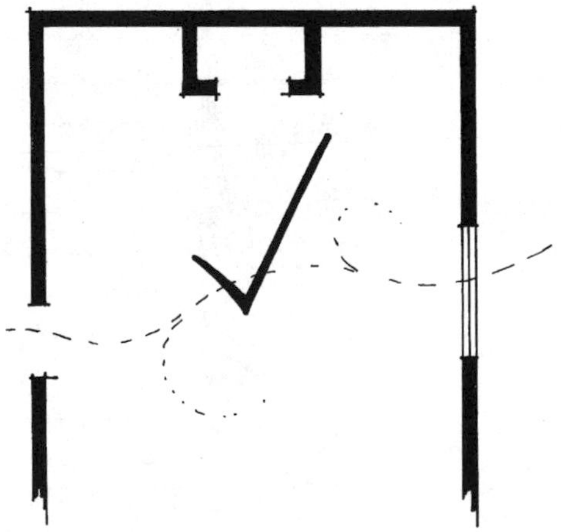

**Figure 15.19**

I always offer residents a bit of advice on the use of their fireplace.

Many folk don't realize that fires should be built up slowly over a period of time. Starting an enormous bonfire in a cold fireplace can crack the brickwork.

# Fireplace without Splayed Sides Internally

Figure 15.20 shows a fireplace without splayed walls and Fig. 15.1 shows the setting out of a fireplace with splayed walls.

**Figure 15.20**

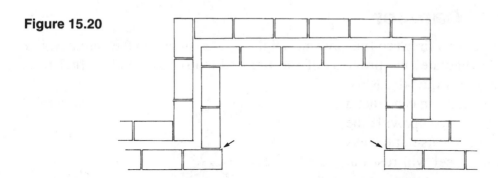

A fireplace without splayed walls is easier to build because the bricks are simply bonded around the perimeter and there is no requirement to cut bricks on an angle. This type of design allows for a bigger area for the fire.

Figure 15.20 shows how I set back the side walls about ⅜" (10 mm) jambs. This allows the back and side walls to be laid in overburned clinker bricks without detracting from the appearance of the fireplace.

Also because the walls are not part of the jambs they can be easily removed and replaced if problems occur. Further, any cracking along the vertical joint would be difficult to detect.

**Figure 15.21**

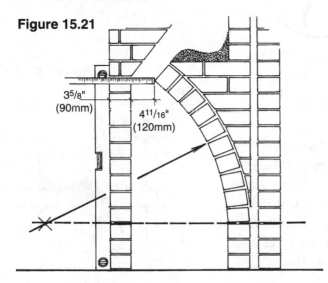

3⅝" (90mm)

4¹¹/₁₆" (120mm)

Figure 15.21 is a sideview of a fireplace showing a rolled or curved backwall. A curved backwall has the added benefit of reflecting heat forward.

When I build this type of fireplace, I cut a template out of masonite or similar plywood to get the curve of the backwall.

Scribe the masonite as per the method shown on diagram left. The fender walls are then constructed. As the end bricks approach the masonite they are marked for cutting. When both fender walls are built the reflection back is freehand built. Because the backwall is similar to an arch, it stands better freely and doesn't require propping.

# Fitting a Damper

A damper in a fireplace flue prevents all the heat escaping up the chimney. I've found it also regulates the progress of the fire and stops the wood or fuel from burning away too quickly. Most engineering works can construct a damper for you if you provide the measurements — use stainless steel as milled steel will rust out in no time.

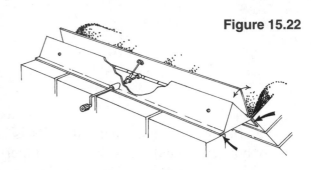

**Figure 15.22**

When fitting a damper take care to fit it loosely. That is, leave a gap of around ⅜" (10 mm) around the steel for expansion.

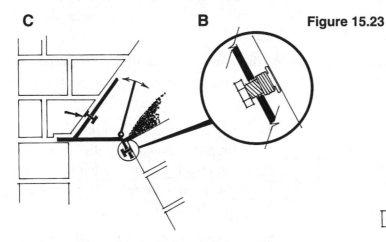

**Figure 15.23**

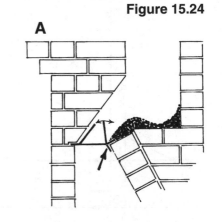

**Figure 15.24**

Figure 15.24 shows the fitting of a damper and the expansion gaps. Figure 15.23C shows the first course brickwork which follows the damper angle and forms the throat and how it is checked out to allow clear space around the lip of the damper edge. It is important that the fireplace jambs are level with the top of the fireback so the damper sits flat.

I use spring-loaded bolts (Fig. 15.23B) to hold the damper in place. They allow expansion from heat but also hold the damper rigidly enough that it can be operated without slopping around.

If the rotary control and handle for the damper should happen to "fall" into line with a perpend, so much the better. If not, it will be necessary to bore a hole through the brick that will receive the control and handle.

This brick is fitted after drilling and not drilled after fitting. I use a short length of ½" (12 mm) round steel rod as a placeholder in the brickwork until the fireplace is complete.

This ensures the alignment of the hole through the brickwork to the damper is not altered by pressure from work on subsequent layers of bricks.

**Figure 15.25**

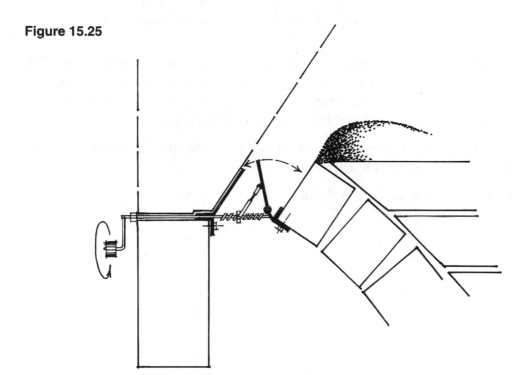

Position the damper in the fireplace in its open portion. Then drape a cleanout rag over the smoke shelf, over the lid of the damper and down into the fireplace. This ensures any droppings of mortar will be caught and can be pulled out with the cleanout rag and will not cause fouling of the operation of the damper.

# Brick Hearth

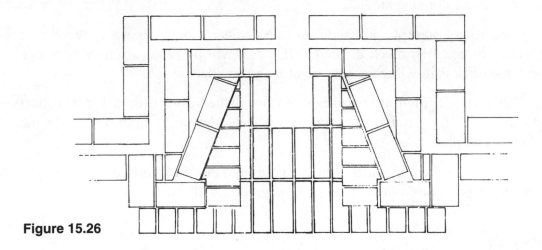

**Figure 15.26**

Lay the inner hearth and hobs inside the fire box first as shown above. The complete inside of the fire box, including the brick hearth, should be round-iron jointed. It is preferable to lay the inner hearth approximately 1" (25 mm) below the outer hearth as this prevents ash and hot material from rolling out.

Mortar for the hearth and fire back should be made of lime mortar, allowing only a half a shovel of cement to each batch.

The lime mortar appears to withstand the heat much better.

# Chimney Stack

Bringing the chimney through the roof requires correct damp-proofing. A copper chimney tray is fixed across the stack above the roof line. The tray may need to be higher on a steeply pitched roof.

The back of the flashing on the chimney should be built up to allow the runoff of water. Depending on the location of the stack, the smoke exit points should protrude above the highest point of the roof or should be located to account for the usual direction of the wind.

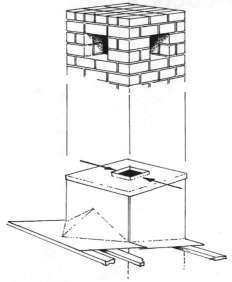

**Figure 15.27**

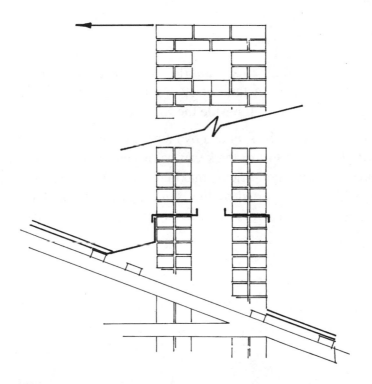

**Figure 15.28**

# Woodbox

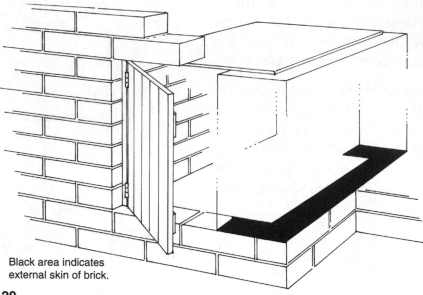

Black area indicates
external skin of brick.

**Figure 15.29**

A wood box is much more easily completed if it is done during the construction of the fireplace. A hole of 12" (300 mm) (1.5 bricks) by seven courses high is left in the rear of the wall beside the fire box.

The front opening can be spanned by an arch matching the front of the fireplace; a piece of Hardiplank ³⁄₁₆" (8 mm) thick can be placed on top of the woodbox; and the top can be filled with common bricks, making sure to back-prop the Hardiplank until the brick roof, which is supported by the Hardiplank, is dry and self-supporting. The brick hearth can also be laid across the woodbox opening and carried inside for a floor.

# Two-Story Fireplace Construction

Including a two-story fireplace in a modern brick house is a common request to bricklayers, especially in colder climates, and I personally think it is a wonderfully warm idea.

It makes a lot of sense in houses that have up and down living areas or bedrooms upstairs and living/family rooms downstairs.

The most important point to incorporating a two-story fireplace in a dwelling is to plan for its construction when the house plans are being drawn up.

The recess on the ground floor needs to be wide enough to ensure there is enough room left for the second flue in the chimney.

This illustration (left) shows 7⅝" (190 mm) brickwork supporting a fireplace on the first floor of a dwelling.

The external skin is taken up to required height but the internal skin is layered two courses lower to support the concrete base which should be at least 7" (170 mm) thick.

The thickness of the base concrete is important because of the weight of bricks it supports. It is necessary to box out a flue exit in the floor, usually 15⅝" × 15⅝" (390 × 390 mm). The inside of the flue exit is parged with a mortar, lime, sand mix. The 15⅝" × 15⅝" flue exit in the concrete allows bricking to carry all the way through so that the concrete base is not exposed.

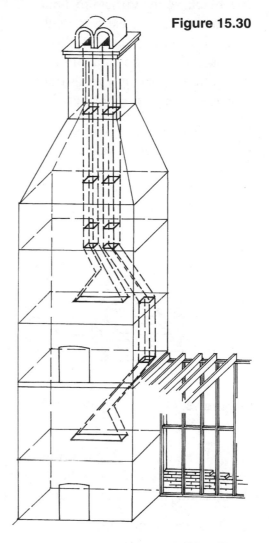

**Figure 15.30**

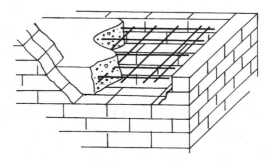

**Figure 15.31**

This is important because the bricks can withstand heat but the concrete base can crack from too much heat and allow seepage of smoke and fumes. This allows parging of the flue to carry through uninterrupted.

The flue is always two skins — the internal skin carries through and the external skin adds additional support to the concrete base of the fireplace.

**Figure 15.32**

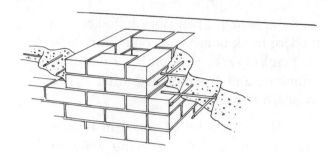

In a two-story fireplace there are always two flues, as bringing two flues into one could create smoking problems. Make sure the withes travel to the top of the chimney opening.

Figure 15.34 shows a straight-through flue. In situations where it is not possible to allow a gathering of the flue, a straight flue is built but would be more viable with a damper fitted.

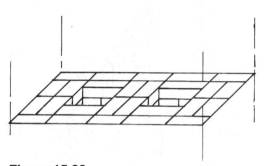

**Figure 15.33**

The purpose of gathering in a flue is twofold. One, to stop wind downdraft bringing smoke back into the dwelling, and two, to slow the rising of the heat so all the warmth from a fireplace doesn't dissipate up the chimney.

However, in buildings with low ceilings, and where it is not acceptable to have a bulky recess external to the house, a straight flue is built.

When we have two or more fireplaces in multistory dwellings it is much better to position the fireplaces over each other. When setting out the flues, build from the bottom. If we have four fireplaces on top of each other we start out at the bottom with three dummy flues and one active one. As we penetrate each floor we brick up to the top of the smoke box and activate the next flue and so on. The breast walls are also able to carry the underflues through to the next fireplace.

**Figure 15.34**

Front view of fireplace
shows combined woodbox
(for rear view refer to
Fig. 15.29)

# Back-to-back fireplace

**Figure 15.35**

A back-to-back fireplace is not too different from a standard fireplace. We still have to build two fire backs, two smoke shelves, two throats, and two gatherings. The only saving that can be made is the withe or the wall between the two gatherings. The withe between the two fireboxes can be single skin because the workings take up so much room there is quite often difficulty fitting them to a specific room size. Offsetting them can save a lot of room (Fig. 15.36).

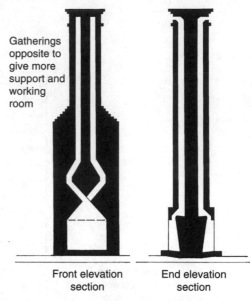

Gatherings opposite to give more support and working room

Front elevation section          End elevation section

Firebox openings

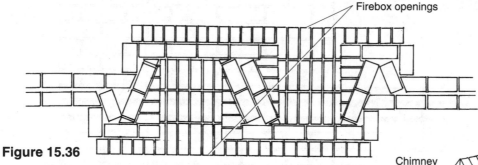

**Figure 15.36**

Be sure to take the withe between the two flues completely to the top of the cut, otherwise wind can blow smoke from one flue down the flue of the other fireplace. Notice the gatherings on the front elevation running in opposite directions. They can also run side by side. When completing the chimney for the fireplace, remember that if you are going to use a roof over the chimney, you must orientate it so as to make the most use of the prevailing breezes.

Chimney stack

**Figure 15.37**

# Fireplace Size

**Figure 15.38**

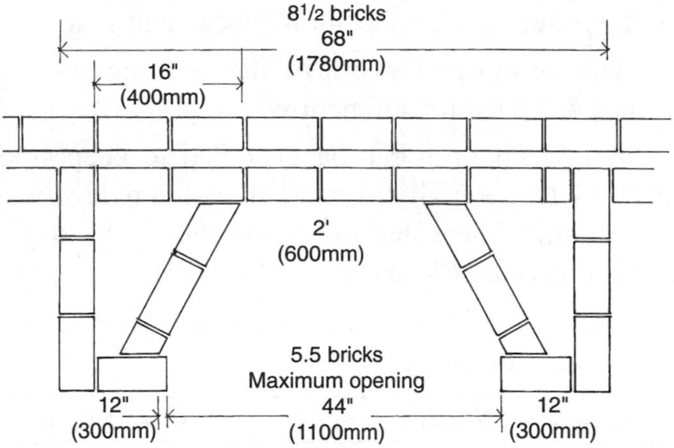

The size of the room does much to determine the size of the fireplace in both technical and aesthetic terms. A huge fireplace in a tiny room will bake the occupants. A tiny fireplace in a ballroom will warm neither body nor soul.

Facings can vary enormously in proportion to openings, to give more or less emphasis to a fireplace as the designer sees fit. Even so there are limits. The firebox cannot be a peephole in a mass of stone, nor can a huge firebox look logical surrounded by a matchbox frame in an otherwise flat wall.

The chart below establishes some conventional relationships of fireplace opening to room size.

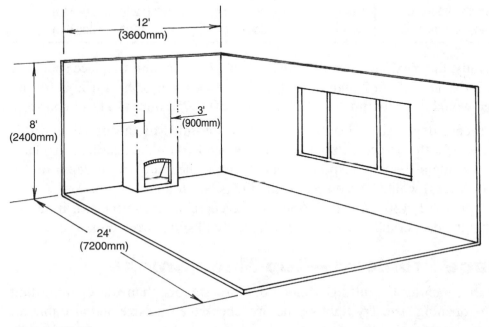

**Figure 15.39**

# Fireplace Opening

No specific rule can be given for fireplace openings because size will vary with the room size, architectural appeal, and availability of wood fuel.

The height and width of the opening should be in proportion so that the fireplace will function efficiently.

A low opening can be expected to keep smoke within the fireplace more effectively than a high one and although a reduction in height will nominally reduce the radiation of heat into the room, this result may be offset in part by raising the entire fireplace a little above the floor level.

Table 15.1 lists the approximate size required to construct fireplaces which will operate efficiently.

**Table 15.1**

| Suggested Width of Fireplace Openings Appropriate to Size of a Room | | |
|---|---|---|
| | Width of Fireplace Openings | |
| Size of Room | Short Wall | Long Wall |
| 10' × 14'8" (3000 × 4400mm) | 2' (600mm) | 2' to 2'8" (600 to 800mm) |
| 12' × 15' (3600 × 5000mm) | 2'4 to 3' (700 to 900mm) | 2'8" to 3' (800 to 900mm) |
| 12' × 18' (3600 × 6000mm) | 2'8 to 3' (800 to 900mm) | 3' to 3'4" (900 to 1000mm) |
| 12' × 21'8" (3600 × 7200mm) | 2'8 to 3' (800 to 900mm) | 3' to 4' (900 to 1200mm) |
| 14'8" × 25'4" (4400 × 8400mm) | 2'8 to 3'4" (800 to 1000mm) | 3'4" to 4' (1000 to 1200mm) |
| 15' × 27' (5000 × 9000mm) | 3' to 3'4" (900 to 1000mm) | 4' to 5' (1200 to 1500mm) |
| 20' × 36' (6000 × 10800mm) | 3'4" to 4' (1000 to 1200mm) | 4' to 6' (1200 to 1800mm) |

Generally, the maximum length of the opening in a medium fireplace should be between 2'3½" and 2'7" (685 and 770 mm). If a brick-on-edge hearth is used, this height range would be approximately 2'2" to 2'6" (650 to 750 mm) (10 to 11 courses).

The recess depth should not be less than one-third of the opening width. However it is clear that to accommodate normal wood fuel on the hearth or in a large grate and to contain the combustion products within the fireplace, a depth of 16" (400 mm) (2 bricks) would be required. If the fire is burnt in a small "basket" grate a shallow depth of 12" (300mm) (1.5 bricks) is acceptable. Nevertheless, it can be expected that the deeper the recess, the less the fire will suffer from cross drafts.

# Fireplace Problems—Too Much Smoke

Test the opening by cutting a piece of fibro smaller than the opening thus reducing the opening size. Try lighting the fire, check the flue size and also that the chimney is past roof ridge.

# 16

# Repairing and Replacing

## Repairing Cracks in Brickwork

Good footings are vitally important to the success of good brickwork. Rarely have I seen brickwork crack without any underlying cause—usually because the footings on which the bricks are laid have moved or cracked. Foundations of heavy clay soils that absorb moisture in rainy seasons and swell up cause most problems for footings.

There is not much point in repairing cracked brickwork if the footings are likely to move again after an extremely dry season or a very wet season. Foundations refer to the soil/substrata on which a structure is built and footings are usually poured concrete or compacted material.

**Figure 16.1**

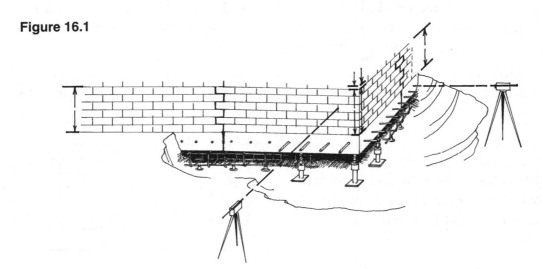

Step 1 is to find where the crack begins. Take a level to the job and site around the construction to determine how far out of plumb the building is. Sinking of footings is common and "growth" of the foundation can cause heaving and displacement of footings.

Trace the fault back to the foundations, dig out all the earth around the footings to a nice solid base, and then jack up the slab or footings until the brickwork realigns. Sometimes it's better to jack a little higher to allow for settling.

What is required is to build a new footing to supplement the existing footing. Use plenty of reinforcing to prevent the new footing from moving apart if it should crack.

**Figure 16.2**

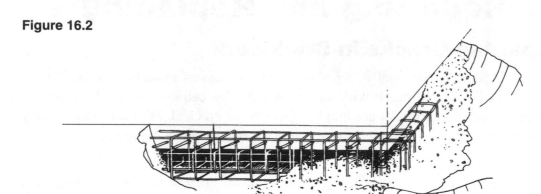

If necessary, drill into the existing footing and hammer in some solid lengths of reinforcing rod, then backfill the entire void with concrete. Don't skimp on the concrete or you could be back in six months to do it again. I don't know how to get the jacks out again afterwards, so be prepared to sacrifice them.

Once you have concreted a new footing you can proceed to cleaning out the grout in the crack line and replacing it with fresh mortar. Replace any bricks that are cracked or broken.

**Figure 16.3**

In cases where the footing is the problem but it is not sufficiently troublesome to be replaced, you can remove the affected bricks (tooth them out carefully) and then refit them with fresh mortar.

In the event the movement is likely to continue but still not sufficiently to require a footing rebuild, tooth out the bricks along the fault line and put in an expansion joint of some reputable material that will allow some vertical movement.

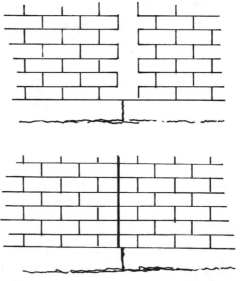

**Figure 16.4**

# Cracking above Doors and Windows

Expansion of lintel bars in very hot weather can contribute to cracking above windows and doors.

When the lintel bar is fitted with mortar up to each end, there is no room for expansion, and as a consequence the expansion forces cause cracking.

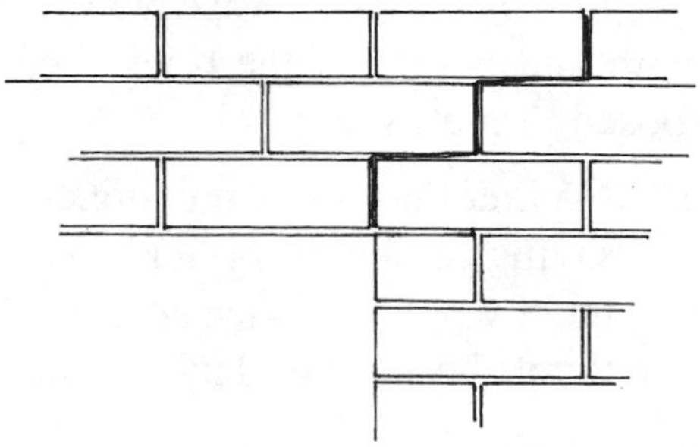

**Figure 16.5**

To overcome this I have often fitted lintel beams onto a piece of DPC and then left about ⅜" (10 mm) gap at either end between the mortar bed on the brick course and the lintel beam.

This setup allows room for the beam to expand and the DPC acts as a slip pad, facilitating any possible movement without affecting the surrounding brickwork.

**Figure 16.6**

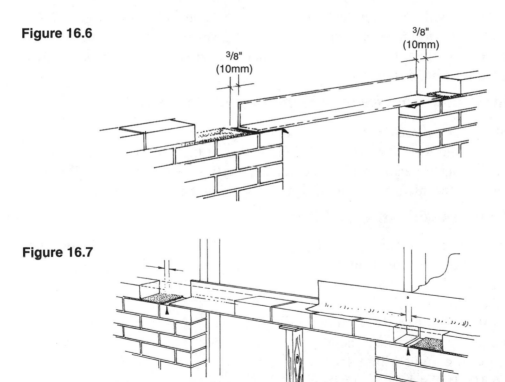

3/8"
(10mm)

3/8"
(10mm)

**Figure 16.7**

Arrows indicate gaps for expansion of the lintel bar.

# Soldier Courses

Soldier courses are where the length and height of the brick face are exposed. Benefits are:

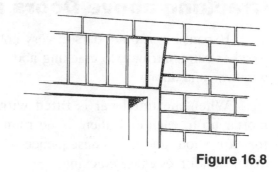

**Figure 16.8**

1. No lintel bars — a big money saving as in cavity brick houses there would be twice as many lintels because of double skin.

2. Especially noticeable in houses that are susceptible to rust, e.g., near the beach; no bars—no rust.

This method is more time-consuming for the bricklayer than fitting lintel bars and stretcher bond over windows and doors, because soldiers take longer to lay. Headers still have to be cut and laid under the window. More thought has to be given to setting out head height of windows and doors.

This height will be the closest course height of one full brick length of 7⅜" (190 mm) added on to the door frame height. This gives the top and bottom of the soldier—the underside of the soldier is our window head height. It is good practice to now mark all sill heights on a gauge rod.

## Keying in Soldier Courses over Window and Doorways

Cut the brick on each end of the window 1" (25 mm) shorter at the bottom of the soldier than the top. The brickwork running up to the cut soldier can now be measured by using a pencil substituted for the vertical perpend joint. (Refer to section on arches for instructions on marking and cutting the brickwork.)

## Soldier Courses without Lintel Bars

Soldier courses can be self-supporting if they are constructed in the right manner. It may require a bit of work with the saw but it is well worth it in the end. Fig. 16.9 shows jamb bricks at the soldier head height, which have been cut out with a saw as per Fig. 16.10. If the bricks are frog bricks, scarf out the opposite side to

**Figure 16.9**

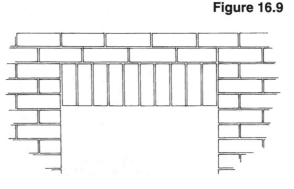

the frog as shown in Fig. 16.11 This will give the soldiers a very strong vertical bond. The same method can be incorporated with Figs. 16.12 and 16.13; only the jamb bricks won't need to be scarfed out on the mitered corners. Also if we are using extruded bricks only the jamb bricks will have to be scarfed out.

**Figure 16.10**

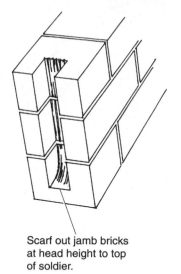

Scarf out jamb bricks at head height to top of soldier.

**Figure 16.11**

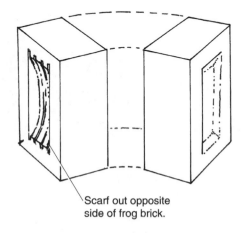

Scarf out opposite side of frog brick.

**Figure 16.12**

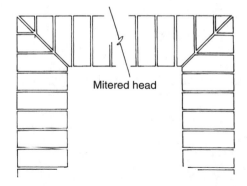

Mitered head

**Figure 16.13**

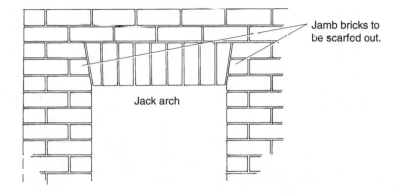

Jamb bricks to be scarfed out.

Jack arch

# Notes

# Notes

# Part III

# Brickwork—Special Effects and Projects

# 17

# Construction of a Brick Kitchen

Kitchens in brick can look quite effective. I've seen very few brick kitchens but they are strong, long lasting, and almost indestructible. The kitchen shown in below was constructed in federation style. It is important however to match the brickwork and the counters. Timber-top counters look good with brickwork. The doors for this kitchen,were also made from timber to finish off the look. In constructing the kitchen, the first course of brickwork has to be rebated for a kickboard. I used two courses of brickwork under the stove and microwave and over the refrigerator and oven. These brick courses were supported by angle iron.

Soldier courses around the top of the cupboards are supported by angle, which gives strength. Be sure to remember where plumbing and electrical services are to be fitted because unlike with timber kitchens, it's much harder to fit these services later if you forget.

**Figure 17.1**

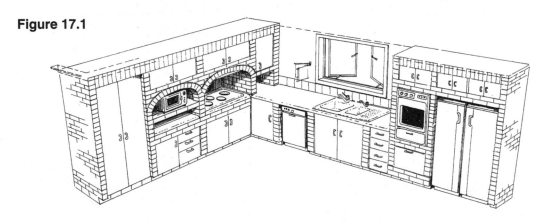

# 18

# Bars

This bar illustrates both an external and an internal curve. It requires all freehand level work. Both curves are made from half bats. On the external curve the half bats are cut to a taper to allow for the front perpends to be laid with a ⅜" (10 mm) joint.

The brick piers at the rear of the bar are to stop two courses below the top of the bar to support a lower shelf.

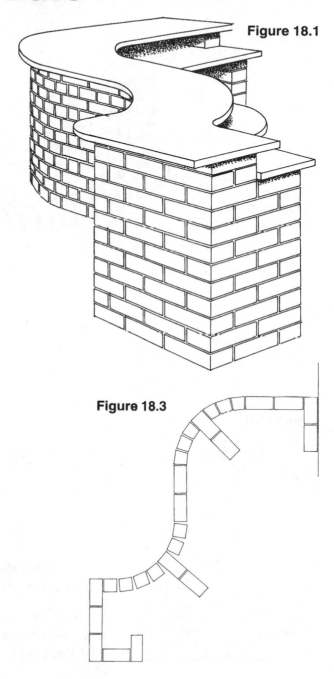

**Figure 18.1**

**Figure 18.2**

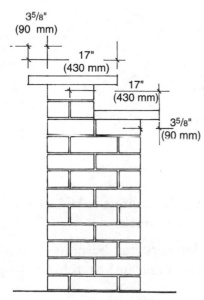

3⅝"
(90 mm)

17"
(430 mm)

17"
(430 mm)

3⅝"
(90 mm)

**Figure 18.3**

# Curved Bar Laid with a Stretcher Bond

The radius shows the bricks are laid with a ⅜" perpend on the front, and pushed hard together on the back. A template can be cut out to make the laying of the brickwork easier (see Figs. 18.5 to 18.7).

**Figure 18.4**

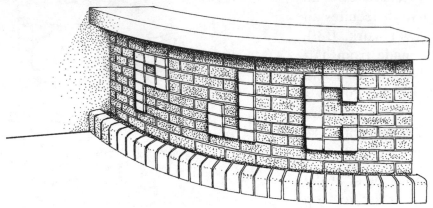

**Figure 18.5**

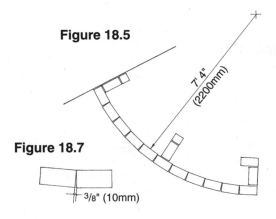

**Figure 18.6**

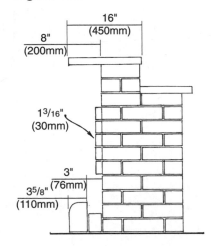

**Figure 18.7**

The sketch shows the footrest is laid as a Soldier course out of bullnose bricks.

Initialing in brickwork can be either protruded (as shown in this diagram) or rebated to give different effects.

# 19
# Barbecues

## Conventional Brick Barbecue—560 Bricks

Almost all residents of the tropics have an outdoor entertainment area, the center of which is usually a barbecue.

Our conventional barbecue has plenty of working room, lots of table space, built-in cupboards for convenience and storage space, and a corbeled chimney for placing utensils. A side fire box means no heat or discomfort for the chef.

**Figure 19.1**

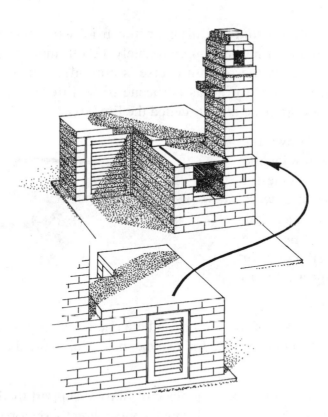

**Figure 19.2**

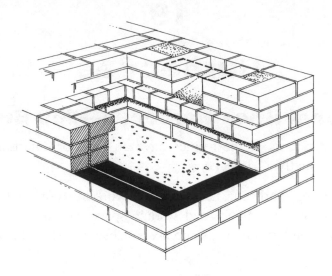

Figure 19.2 shows that building up the brickwork to the fire box, and protruding out a course of headers approximately 1⅜"(35 mm) at the correct height to carry the grate (usually the seventh course is protruded) allows easy removal of the grate. This also allows for a gas barbecue to be fitted later if required. The remaining brickwork up to the plate is called the fire box.

Figure 19.3 shows the setting out of the brickwork, allowing 1.5 bricks wide for the chimney. The brickwork up to two courses underneath the grate should be filled with rubble and either paved or concreted, leaving a table to remove the falling ash.

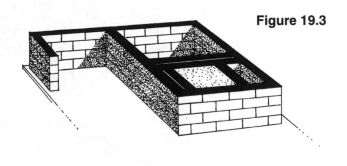

**Figure 19.3**

During construction of the fire box, a hole one brick long by two courses high will be left out on the chimney side for a smoke box. Make sure the mortar inside the smoke box is sloped.

The dotted line in Fig. 19.2 is a piece of fibro to support the brickwork across the opening of the smoke box. The chimney course directly on top of the smoke box should be raked out approximately 1" (25 mm) to allow the ⅜" (10 mm) plate to slip underneath, sealing tightly into the mortar bed joint.

# Alternative Barbecue Design

This barbecue was designed for the television program Burke's Backyard.

## Barbecue No. 1 (without table)

The brick courses are laid as per drawing, for six courses, we then stick out a brick to carry the grate. We do this because any steel bearers laid in the mortar bed joints to carry the grate will create cracking of the brickwork as the heat expands the steel. A piece of fibro is laid at the back of the fire box to easily cover the hole, and as a base for the smoke box. A half-brick wide by two courses high is left out, and the mortar is sloped inside the smoke box to allow easy exit of smoke. Now another piece of fibro 8" long x 3" wide (200 × 70 mm), is placed over the hole, and a pier is constructed to a height, generally over eye height.

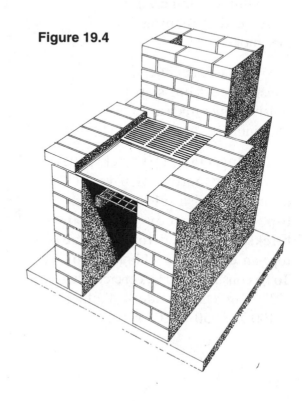

**Figure 19.4**

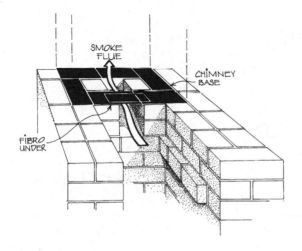

**Figure 19.5**

This brick barbecue has been set out to suit the universal barbecue plates and grills easily obtained from "Barbecues Galore." Also it has been constructed out of double brick for the handyman's benefit, as the extra skin of bricks contains the heat inside the fire box, and thus any person touching the outside of the barbecue will not get burned.

The only brick that has to be cut is a brick which could be broken in half with a hammer, used to form the opening for the smoke box.

**Figure 19.6**

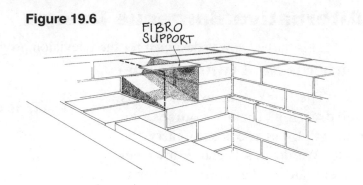

This barbecue was constructed from start to finish by two tradesmen in 45 minutes.

**Figure 19.7**

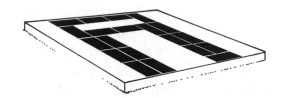

**Materials** 210 bricks, 2½ barrows of mortar, 21 pavers (if desired to finish off the barbecue top, they also cover holes in bricks, 15 needed for barbecue top, an extra 6 for chimney top). To be constructed on a concrete slab 4'8" long x 3'6" wide × 4" thick (1400 × 1050 × 100 mm).

**Figure 19.8**

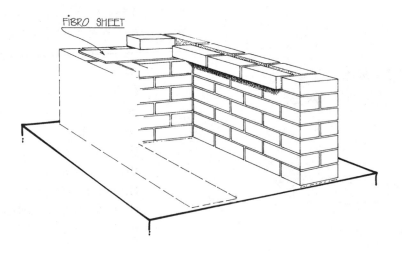

## Barbecue No. 2 (with table)

Simpler in construction, consisting of two walls laid parallel to each other, with a brick pier 1½ bricks square constructed between the two walls. The pier forms the chimney as in barbecue no. 1.

It has for the tabletop, a precast concrete top 2' × 3' (600 × 900 mm) which is laid on a bed of mortar (this can be tiled later if desired).

**Figure 19.9**

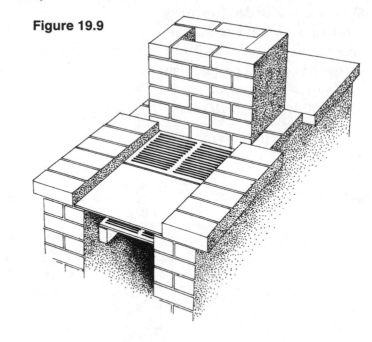

This barbecue was constructed by two tradesmen bricklayers in 55 minutes.

**Figure 19.10**

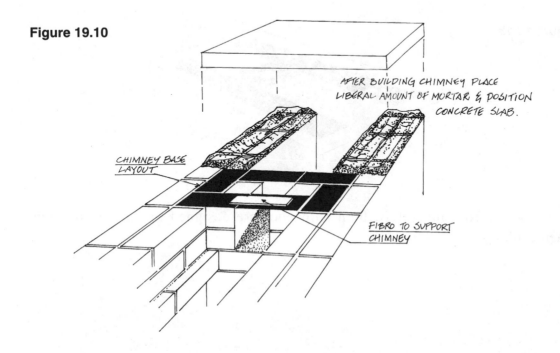

AFTER BUILDING CHIMNEY PLACE LIBERAL AMOUNT OF MORTAR & POSITION CONCRETE SLAB.

CHIMNEY BASE LAYOUT

FIBRO TO SUPPORT CHIMNEY

**Materials** Approximately 300 bricks, just under 4 barrows of mortar, 21 pavers (as for barbecue no. 1 if desired). To be constructed upon a concrete slab 6'4" long × 3'6" wide × 4" thick (1900 × 1050 × 100 mm).

**Figure 19.11**

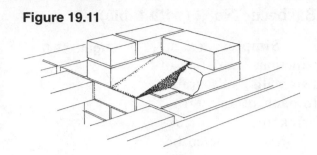

**Figure 19.12**

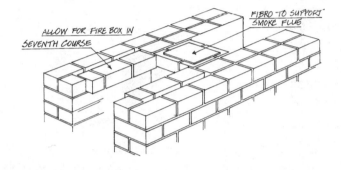

**Figure 19.13**

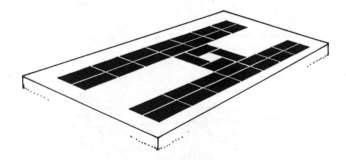

For comfort and safety, both barbecues are designed to stand at the side when cooking.

**Figure 19.14**

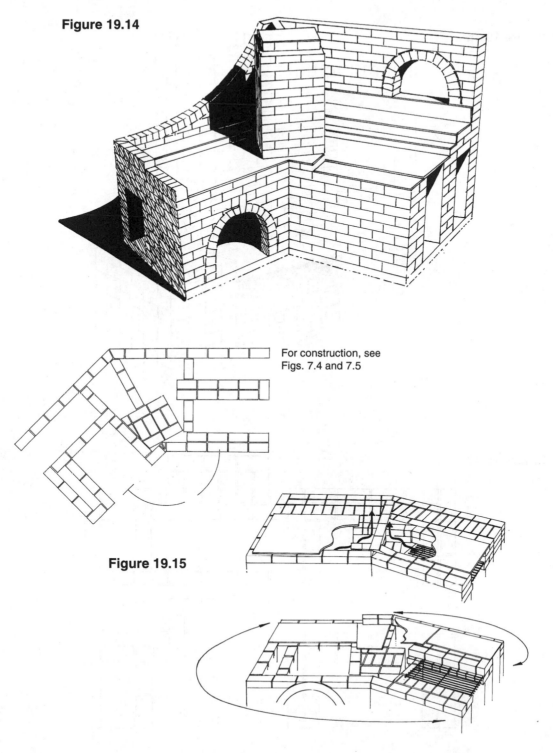

For construction, see
Figs. 7.4 and 7.5

**Figure 19.15**

**Figure 19.16**

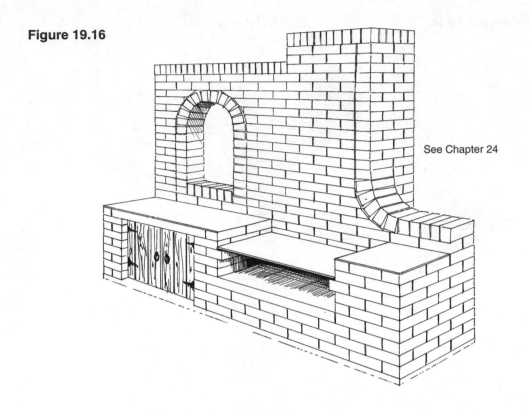

See Chapter 24

**Figure 19.17**

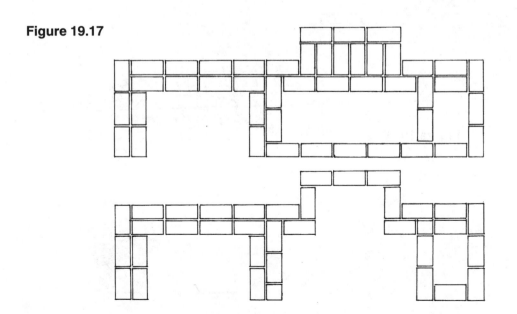

**Figure 19.18**

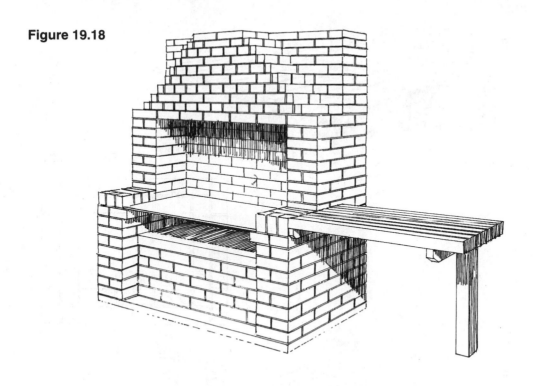

**Figure 19.19**

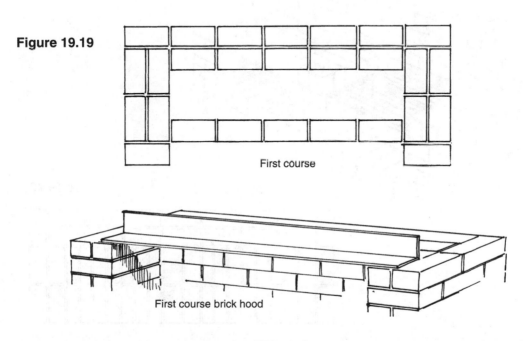

First course

First course brick hood

**Figure 19.20**

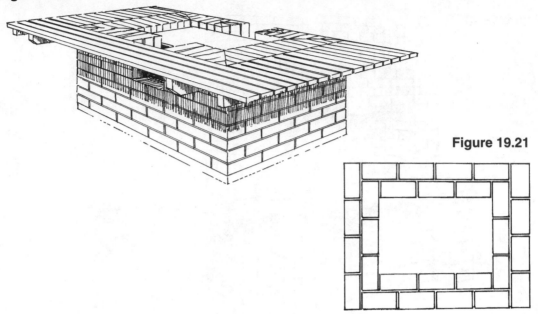

**Figure 19.21**

**Figure 19.22**

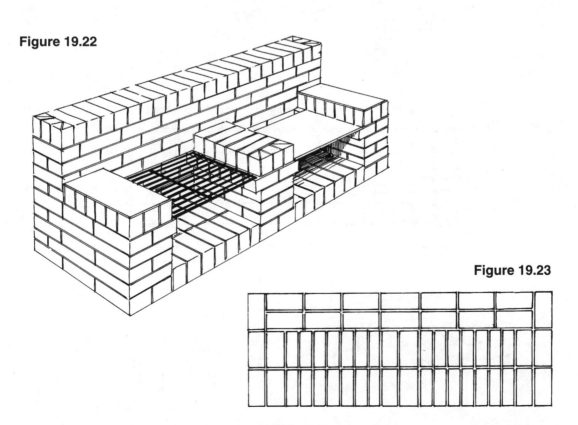

**Figure 19.23**

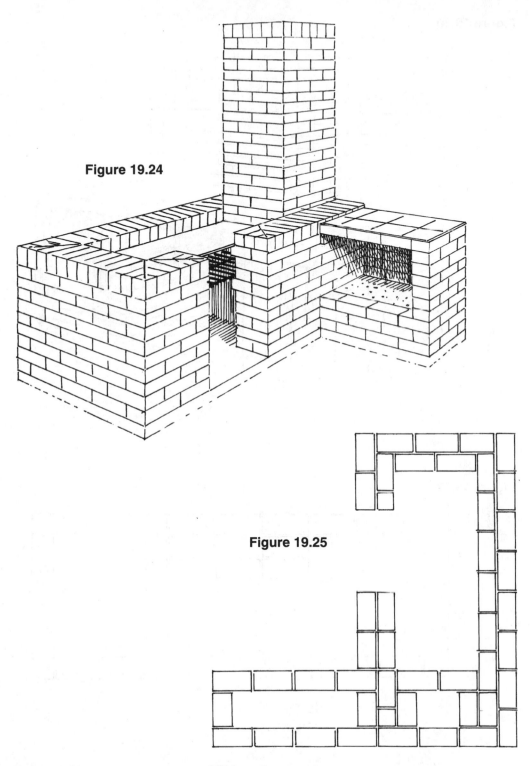

**Figure 19.24**

**Figure 19.25**

**Figure 19.26**

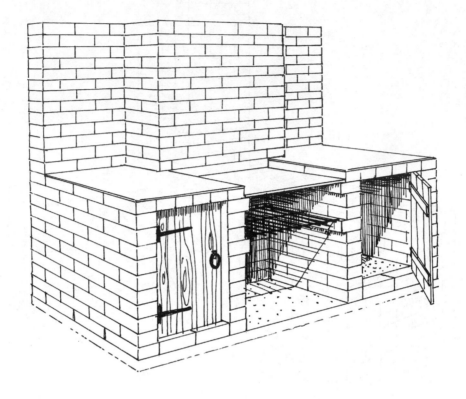

**Figure 19.27**

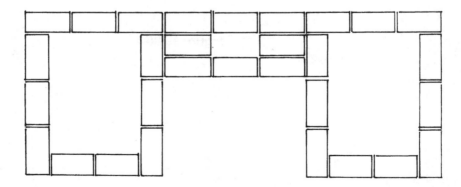

**Figure 19.28**

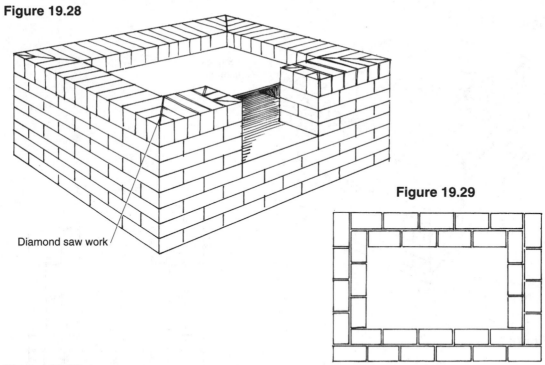

Diamond saw work

**Figure 19.29**

**Figure 19.30**

Diamond saw work

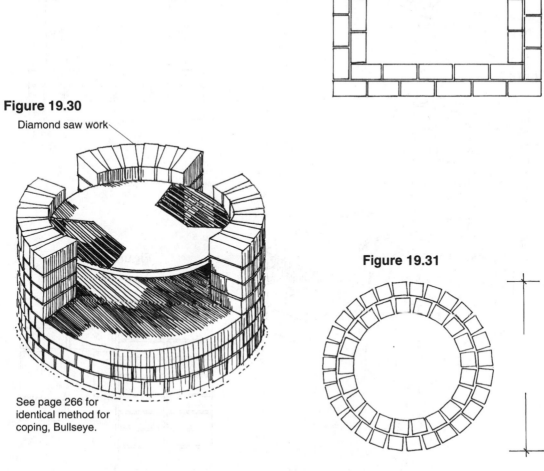

See page 266 for
identical method for
coping, Bullseye.

**Figure 19.31**

**Figure 19.32**

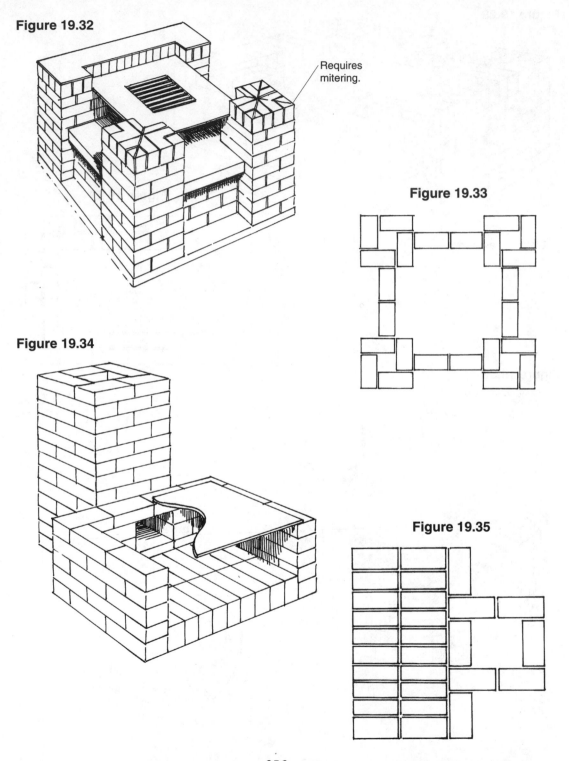

Requires mitering.

**Figure 19.33**

**Figure 19.34**

**Figure 19.35**

# 20

# Construction of Combined Brick Letterbox and Planterbox

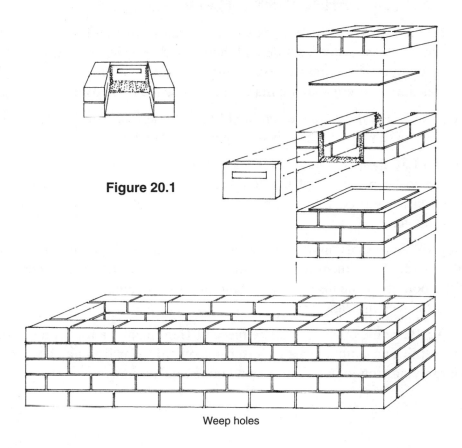

**Figure 20.1**

Weep holes

Not all the work done by a bricklayer is glamorous. On almost every house I've built there has been the requirement to construct a letterbox or planterbox to match the new brick home.

The same skill and attention to details is required in building a letterbox as in building the house, mainly because it stands out on the street where it is most susceptible to scrutiny and damage.

# Materials Required:

1. Cut two pieces of Hardiplank 15" (375 mm) square—these are the top and bottom of the letterbox.

2. Two "Inserts" to suit brickwork or blockwork can be purchased from a local hardware. These are the front and back of the letterbox.

3. You will need one and a half barrows of mortar and a concrete slab 7' × 2' × 4" (2.1 m × 0.6 m × 100 mm). For the slab use a piece of F62 gauge reinforcing mesh 6'8" (2 m) long by 20" (500 mm) wide.

**Step 1** Lay the first five courses as illustrated, making sure to lay the dividing brick skin between the letterbox side and the planterbox side, leaving weep holes in the bottom course of the planterbox. Choosing a 2 × 2 brick letterbox fits the conventional size of inserts that are made one brick wide by two courses high.

**Step 2** Lay the three courses on top of the letterbox-end of the planter, placing the 15" square base with the smooth side up on the brick base.

**Step 3** Lay the two end courses leaving a one-brick hole front and back, and making sure to keep the top of the courses level as in all brick construction. Any unevenness in the top course of bricks will upset the fitting of the inserts and the placing of the lid.

**Step 4** Thoroughly mortar up both inserts and the edges of the bricks into which they slide. Slide them into position. Gently sloping the excess mortar on the back of the bottom of the inserts with a pointing trowel helps keep the inserts in and adds to the finished appearance.

**Step 5** Place the other 15" square piece of Hardiplank on top with smooth side down. Lay the cap course on the letterbox on this Hardiplank. For improved strength, run your iron jointing tool around outside of inserts.

# 21

# Honeycomb Brickwork

This type of brickwork is often referred to as "breeze-way" brickwork or "hit-and-miss" brickwork. The reasons for both names are obvious. I've always preferred to call it honeycomb brickwork.

It has a variety of applications in letting through the sunlight and the cool breezes, but it also lets in rain on windy days. It has not escaped the author either that vermin can get through this type of brick construction. It is always helpful to be aware of these "tricks" because often home owners give little thought to the functionality of a structure. Usually they build or buy for cosmetic reasons. Your helpful advice—which could save them a lot of time, money, and possible disappointment later—will be appreciated, and could see you win their next job over a rival bricklayer.

**Figure 21.1**

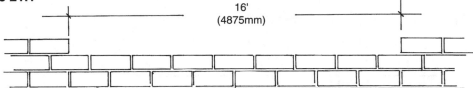

16'
(4875mm)

1. Distance required $\qquad$ $X$ mm

2. Minus 1 hole $\qquad$ $X - 3"$ (75 mm)

3. Divide by length of brick plus 1 hole $\qquad$ $\dfrac{X - 3"}{10\frac{5}{8}" \ (265 \ mm)}$

   $[7\frac{5}{8}" + 3" \ (190 + 75mm)]$

4. (a) If it will divide equally, no problem, use a 3" block of wood for spaces.

   (b) If it does not divide equally

   1.   16'

   2.   16' − 3" = 15'9"

   3.   $\dfrac{15'9"}{10.625"} = 17\frac{3}{4}$ spaces

## Part III: Bricklaying—Special Effects and Projects

This leaves us 17 and ¾ of a brick, i.e., 0.75 of 10.625" which equals 8" (200 mm).

We have to make up this 8" by making each hole slightly wider. Dividing this enlargement over the 17 spaces gives an extra ½" (12 mm) per space. The length of the hole is now 3½" (90 mm).

**Figure 21.2**

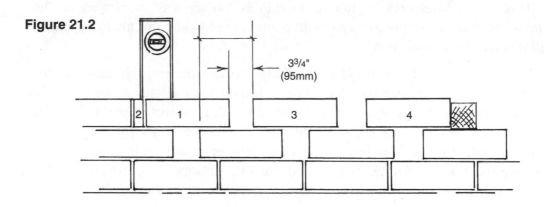

**Figure 21.3**

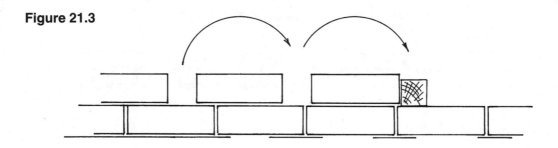

Cut your filler block of wood accordingly.

# 22

# Glass Bricks

First, check the length and height of the opening to be bricked up, because they have nil initial rate of absorption (IRA), they have a tendency to wobble and sink. Laying any blocks over six courses is virtually impossible without the spacers in Fig. 22.1. This means that there can be no opening or the bed or perp joints. Previously, I have used lattice in the bed and perp joints. By varnishing or painting the required color the joinery can be left between the bricks.

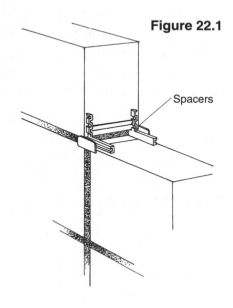

**Figure 22.1**

Spacers

**Figure 22.2**

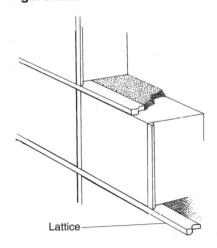

Lattice

It not only looks good but enables us to lay them to the required height (Fig. 22.2) It is even more difficult laying them in a curve. Laying quad in the bed joints makes the job easier (Fig. 22.3).

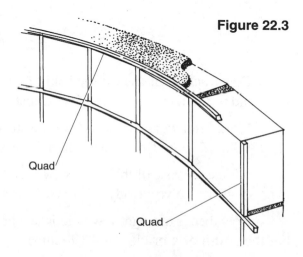

**Figure 22.3**

Quad

Quad

# 23

# Brick Steps

This is a little difficult to explain fully but the diagrams are quite self-explanatory. The construction of brick steps—internal or external—is similar. First, determine to what height the steps have to rise. Determine a gauge to take you from this point to the ground level. Remember, however, that the gauge can be varied and often the height of the floor can also be "moved" to accommodate.

If a concrete slab or path is to lead up to the first course of stairs then a working gauge becomes easier. Try to maintain a minimum distance of 5¼" (130 mm) from the height of the path to the top of the first tread of 192 mm.

**Figure 23.1**

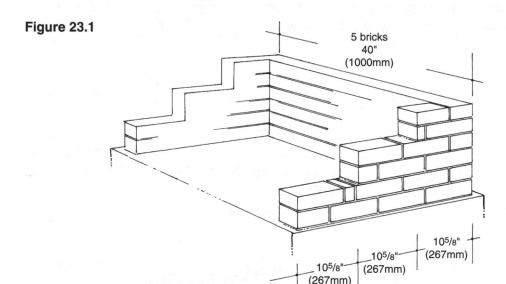

Construction of the external skin first is done to allow easy access to the back wall, and to the risers on either side. Construct risers out of two courses of bricks.

Mark brickwork on the side of the stringers under the tread, called the "going," 12" (300 mm) long. Brick treads are laid on top and protrude over 1" (25 mm).

For the purpose of this exercise we need to look carefully at a brick shape and size in perspective. We've already referred to stretchers, headers, soldiers, and bricks on edge.

How then is the right way to look at a brick? The height of a brick is 2¼" (57 mm), but the width of a brick is 3⅝" (90 mm).

This leaves only one part to be measured and that is the length which is, of course, 7⅜" (190 mm). When laying treads they are placed so the face you see is the length by the height, 7⅜" × 2¼" (190 × 57 mm).

They butt up at right angles to bricks laid on their edge so the face you see is 2¼" × 7⅜" (57 × 190 mm).

Tread brick 7⅜" (190 mm) + ⅜" (10 mm) + 2¼" (57 mm) + ⅜" (10 mm) = 10⅝" (267 mm)

| Brick | Perpend Joint | Brick on Edge | Perpend Joint | = 9⅝" (242 mm) |
|---|---|---|---|---|
| | | | | [Going (1" Overhang)] |

**Figure 23.2**

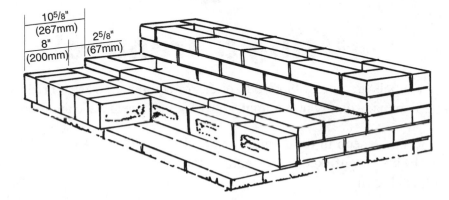

Figure 23.3, shows an alternative set of brick steps constructed on a 4 × 4 brick base.

This method of construction offers the advantages of being able to keep the step bricks flush with risers and also enables the brickwork to be bonded on the corners to eliminate straight joints. The width of the treads in this case is one brick plus two joints and the width of a brick face.

**Figure 23.3**

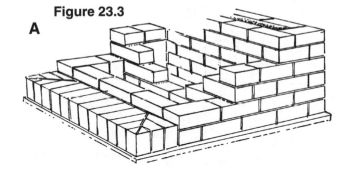

**Figure 23.4**

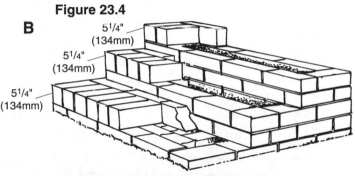

Plan for these steps are four bricks length + four bricks width.

The tread bricks will work neatly 12 across. (Remember to deduct one less mortar joint!)

In the case of a set of high stairs (more than three steps) it is important to work your steps down from the top so you can build the internal walls without difficulty. If you start from the lowest end, by the time you reach the third step it will be nearly impossible to get bricks into position for the vertical walls. An alternative is to pour a slab and build up from the slab.

# 24

# Using a Trammel

Trammels are one of those odd pieces of equipment that no bricklayer should be without. I've made up my own trammel out of some pipe and g-clamps and a pivot joint.

As can be seen in the diagram below, I have a Nikko pen affixed to the end of the multiadjustable jig.

**Figure 24.1**

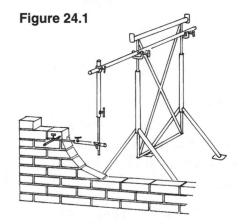

**Figure 24.2**

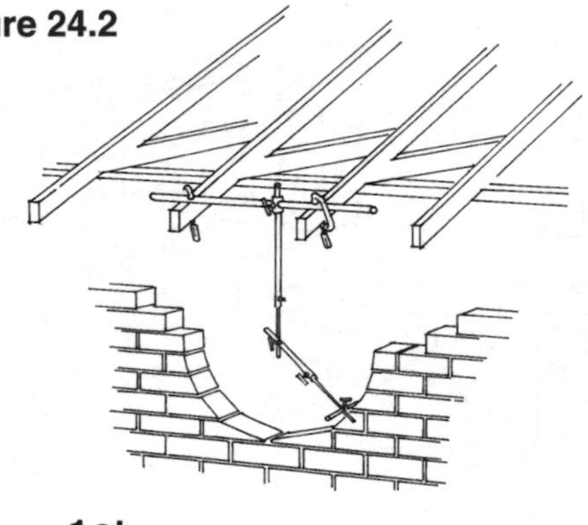

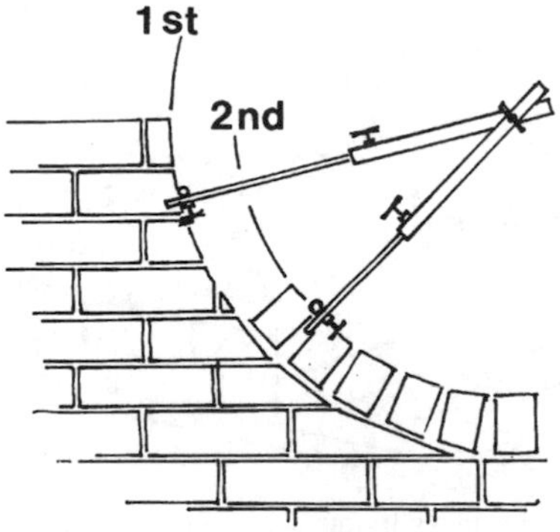

**1 st**

**2nd**

**Figure 24.3**

This allows me to scribe inverted arches, round windows, and bullseyes.

The trammel can also be adjusted back 4" (100 mm) to allow scribing around the inside diameter of the circle or arch.

The trammel remains in place and as each brick is tried in position the cut mark is made with the Nikko pen.

It is simple and effective and saves building templates.

# 25

# Bricking in a Round Window

Bricking in a round window is not as difficult as it seems. First, you must find the center by marking the diameter of the window on a level. Move the level up and across the window until the marks are at the outside edges. Mark the center where the marks meet.

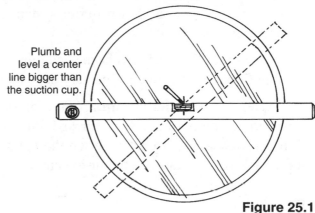

Plumb and level a center line bigger than the suction cup.

**Figure 25.1**

**Figure 25.2**

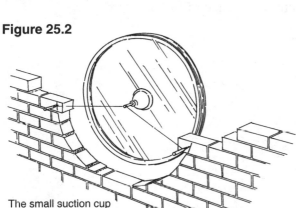

The small suction cup is to hold the line holder timber level.

Next we need a cheap suction cup plunger like plumbers use to clear sinks. Measure the diameter of the suction cup and place those marks on the plumb and level marks, (horizontal and vertical). Fit the suction cup to the window and place a level across the window. Mark the handle of the plunger and then cut off the plunger handle flush with the face of the brick work. Attach a string to the handle and with a waterproof pen as shown in Fig. 25.1, and allowing the required distance for the length of the voussoir and the mortar joint, the brick work outline can now be scribed with ease. [The additional distance for the length of the voussoir and the mortar joint would usually be 4" (100 mm).]

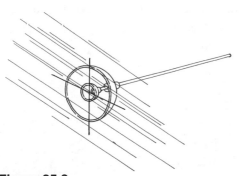

**Figure 25.3**

# 26

# Bullseye Arch

Find the circumference of the outside circle you are working to. Divide this by the appropriate gauge as shown in our system. This will give the number of bricks around the circumference.

Sometimes the thickness of the bed joint between the bricks has to be varied to make the fit better. It is often necessary depending on the type of bricks used.

After finding the circumference of the inside circle, divide the inside circumference by the number of bricks in the outside circle. Deduct ⅜" (10 mm) off this measurement for a joint between the voissours. This will give the size of each cut for each voissour around the inside circumference of the bullseye.

**Figure 26.1**

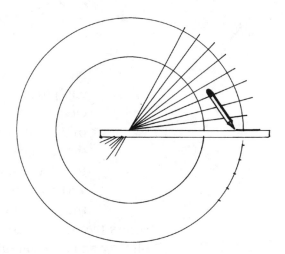

For the exercise we are going to undertake below, I have used an outside circumference of 100½" (2550 mm) which is arrived at by multiplying the diameter of 32" (812 mm) (in this case) by pi which is approximately 3.14.

$X$ = outer circle

$Z$ = inner circle

$G$ = outer gauge

$IG$ = inner gauge

$L$ = length of brick

$Y$ = number of bricks

To find the inner circumference we need to find the diameter:

$$Z = \frac{(X)}{3.14} - [(2 \times L) \times 3.14]$$

To find the number of bricks around this circumference:

$Y = X$ divided by $G$

To determine the inside gauge:

$IG = Z$ divided by $Y$

**Example:**    $X = 100.5"$

$Z = ?$

$G = 2\frac{5}{8}"$

$IG = ?$

$L = 7\frac{5}{8}"$

$Y = ?$

$$Z = \left( \frac{100.5"}{3.14} \right) - 16" \times 3.14$$

$$Z = 32" - 16" = 16" \times 3.14 = 50\frac{1}{4}"$$

To determine the number of bricks: $X$ divided by $G$

100.5" divided by $2\frac{5}{8}"$ = 38 bricks

The inside gauge equals the inner circumference divided by the number of bricks.

In this case 50¼" divided by the number of bricks, 38, equals 1⅓" (33 mm).

But remember ⅜" (10 mm) has to be deducted for our perpend joint.

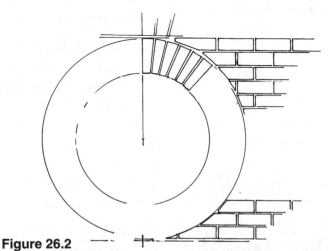

**Figure 26.2**

# 27
# Spiral Piers

This is the system for constructing a spiral pier of 12" × 12" (300 × 300 mm) or one and a half bricks by one and a half bricks. Here I will give the instructions for turning our spiral a full circumference of a circle, 360 degrees. From these measurements, spirals of various degrees can be built. (Figure 27.1 shows 180°.)

Firstly find the height of the pier to determine the number of courses. This is divided by the gauge. Divide this number of bricks into the circumference of the pier.

We need to find :

$$\text{Hypotenuse} = \sqrt{(L^2 + B^2)} = \sqrt{(12" \times 12" + 12" \times 12")}$$
$$= \sqrt{288} = 17"$$

Find circumference: Diameter × π (3.14)

$$= .17" \text{ m} \times 3.14 = 53.38"$$

Height of pier, e.g., closest course to 10' = 10' (3.058 m)

10' (3.058m) = 45 courses at 2⅝" (67 mm) gauge

Now divide the circumference (53.38") by 45 courses = 1.186"/2 = 0.6" for a 180-degree spiral turn.

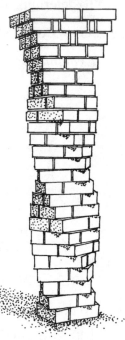

**Figure 27.1**

**Figure 27.2**

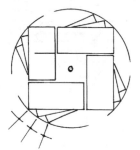

## Construction

Cut a piece of board 12" × 12". Drill a hole through the center to fit over a piece of threaded cyclone rod. After laying the first course, spread a mortar bed on top of the course, now offset the first brick the required distance, e.g., 1.18". The spiral board can be moved down on top of the brick and the other three bricks are pushed in underneath to the correct turn.

The board can be pushed hard on top of the pier and used to level the whole job. Tap the bricks in so they are neat and tidy.

**Figure 27.3**

# 28

# Inverted Arch

The inverted or upside-down arch is a decorative design commonly used in fences, dividing walls between commercial properties, and outside entertainment areas as a feature wall. I've seen them fitted with wagon wheels and they look rustic and not unattractive.

**Figure 28.1**

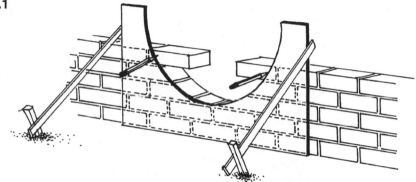

Place a timber template cut to the preferred shape beside the brick wall at the correct height.

Doing this will allow you to scribe the shape on the brick as each course is laid.

Following your stringline for each course, place the brick in position and draw around it with a pencil.

Having cut and laid the brickwork to the finished height, the voissours can now be laid.

## Internal Templates

Lay the brickwork up to the under side of the arch. Lay a voissour and place the template on the voissour, bracing the template as shown.

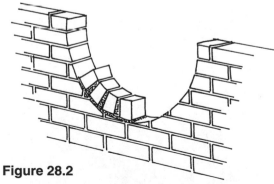

**Figure 28.2**

This method can be used in the fitting and bricking of round windows, as shown in the bricking in of a wagon wheel (Fig. 28.4).

A timber template can be cut 4" (100 mm) larger to allow for the brickwork up to the voissours to be cut in, giving a base to lay the voissour.

(Refer to Fig. 28.2.)

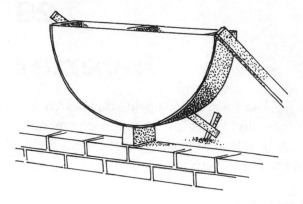

**Figure 28.3**

**Figure 28.4**

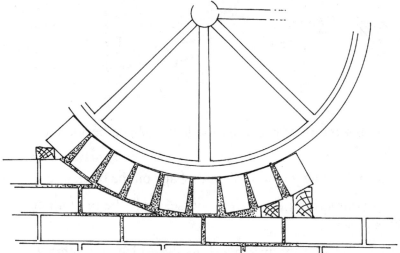

# 29

# Construction of a Wood-Fuel Hot-Water System in Brick

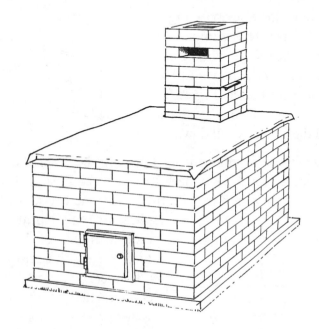

**Figure 29.1**

Construction is similar to building a barbecue. Lay the outside perimeter walls up to four courses making sure to leave a hole for the door. Allow one course of bricks on the floor because placing a fire directly on a concrete slab is not desirable as over time it will cause the slab to crack.

To utilize all the heat in the fire box I like to batter the two side walls and the chimney end wall (Fig. 29.2). This has the effect of reflecting the heat back and preventing loss of heat up the flue.

**Figure 29.2**

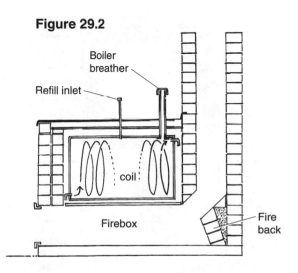

Boiler breather

Refill inlet

coil

Firebox

Fire back

In all internal construction I like to use full lime mortar with a half a shovel of cement. This is different than mortar, where we use a half a shovel of lime and half a bag of cement. Lime mortar can handle the heat better than cement mortar. Another reason for using the cement in the lime mortar is to stop ants and pests from burrowing into the mortar and in time carrying the mortar away particle by particle. The inside skin is left down one course lower during construction so the lid can sit flush with the top of the outside brickwork.

When constructing the flue don't forget to allow a slot for a damper. This is important in preserving heat and keeps the wood from burning away too rapidly. Take notice of the bond where the damper is fitted in (Fig 29.3). If the slot were below the full brick, after a while the brick would eventually fall down.

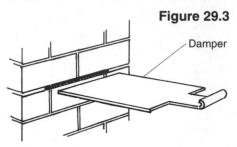

**Figure 29.3**

Damper

The bottom arrow on the coil is the inlet pipe for the water, the outlet is at the top. Be sure to allow openings for these outlets. I use a piece of steel directly over the fire (like a barbecue plate) on which the drum is positioned. Use a ⅜" (10 mm) or ½" (12 mm) steel plate and it will also protect the life of the drum. (Fig. 29.4). In this case the drum was a 44 gallon (200 liter) oil drum with a lid cut out. The coil we fitted was from another older hot-water system no longer used. The top of the drum is welded back in place after the coil is inserted. The larger pipe protruding from the top is a vent pipe. As the water in the drum boils, pressure builds up and is released through the vent. The smaller pipe is to refill the drum. This type of heater can also be used without a coil and just the drum filled with water.

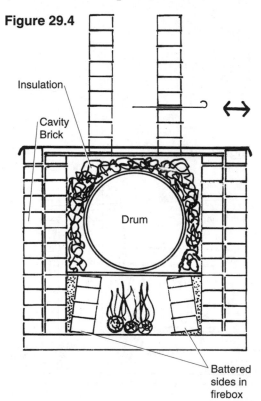

**Figure 29.4**

Insulation

Cavity Brick

Drum

Battered sides in firebox

We then pack around and on top of the drum with fireproof wall insulation. A sheet of galvanized iron for a roof will finish the hot-water system neatly and provide a weatherproof cover.

# 30

# Brick Compost Bin Construction

First excavate, allowing enough room to lay screen blocks or grass pavers inside the footings. The pavers prevent the removal of the dirt floor, but also allow worms access to the dirt below (which is the major factor in compost) (Fig. 30.2). Now run the brickwork up three courses making sure to pre-slot the jamb bricks that hold the compressed fibro door. Then run a lintel over the opening and brick the remaining four courses. Notice the second last course is rebated to hold the fibro lid.

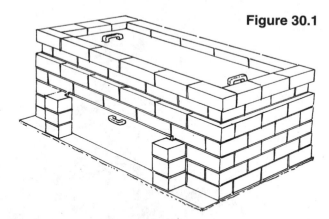

**Figure 30.1**

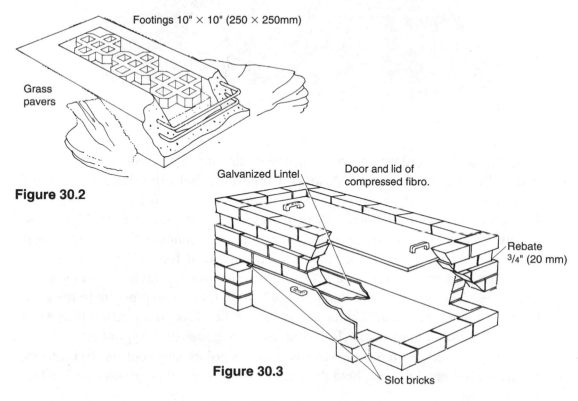

Footings 10" × 10" (250 × 250mm)

Grass pavers

**Figure 30.2**

Galvanized Lintel

Door and lid of compressed fibro.

Rebate 3/4" (20 mm)

Slot bricks

**Figure 30.3**

# 31

# Spiral Arch Waterfall

**Figure 31.1**

As a qualified tradesman and an active contractor, I was asked to have some input into the 5th Australasian Masonry Conference held in Gladstone, Central Queensland. I decided to construct something different to commemorate the conference. I had very little time, only two weeks before the conference. Myself and two of my most capable tradesmen built the spiral arch monument. My helpers built the spiral pier. Reinforcing in the spiral arch consisted of four ⅝" (Y16) rods, a ⅝" (16 mm) threaded rod as well as a ½"(12 mm) water pipe to circle the water to the top of the pier. The ⅝" rods were tied into a ⁵⁄₁₆" (8 mm) coil and pre-bent to the exact radius. Most important is the ⅝" threaded rod which fits through the center (Fig 31.2). I previously marked the rod every four courses at the gauge of the spiral arch.

I then screwed on nuts and washers at these points and bent the rod into the exact curve. Tightened the nut, held the brickwork firm. We then grouted and rodded concrete around the steel making sure to fill completely. I used G clamps to hold the bricks together, especially as it is 7⅝" (190 mm) around the arch. The hardest part was the crown to the spiral arch, where I think my two helpers put a few extra words in the dictionary. After the first day we had the spiral arch sitting on the spiral pier. The next day we finished off the top of the pier and worked out the amount of courses to have the bricks protruding out to take the water from the top of the fountain, directly into the groove chased at the back of the spiral arch. We then left the brickwork cure for a

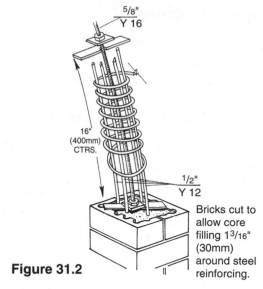

**Figure 31.2**

5/8" Y 16

16" (400mm) CTRS.

1/2" Y 12

Bricks cut to allow core filling 1³/₁₆" (30mm) around steel reinforcing.

week before coming back to cut the rebate around the arch to allow the bricks to sit in. The bricks themselves have a slot about 1³/₁₆" (30 mm) wide and 1⅝" (40 mm) deep to carry the water down to the catchment. We grouted the catchment and painted the entire monument with damit 800.

I would like to thank the Central Queensland University, especially the Engineering and Science departments for their support. Also Manicka Dhanasekar who without his persistence I would not have been able to build the monument. I was also able to witness some of the time put in by these lecturers. I received many a phone call on the weekend and meetings in his office at night, after an already long day.

Perhaps it is the symmetrical or rectangular shape of bricks that has led engineers to design and bricklayers to build in rectangular construction. Bricklaying, however, can be an expression of an individual's own imagination and creativity.

Brickwork is extensively used as a structural and nonstructural architectural building material, and even though rectangular construction is the most popular, horizontal and vertical curves and spiral shapes of construction are also possible. The latter form of construction requires good engineering judgment, architectural sense, and highly professional craftsperson skill.

The water fountain is an ornamental piece designed to capture attention and imagination, to challenge young and training bricklayers to think outside the block, and to explore ways to make their work interesting and appreciated.

# Specifications

The brickwork construction on display may not be exactly in accordance with the plans and sketches provided. During on-site construction the bricklayer needs to be a little versatile and be able to improvise for the sake of design.

## Clay Block Spiral Column (Fig. 31.4A)

Height 16" (4000 mm). With clay block gauge of 3⅜"(85 mm) there are 47 courses. Minus three courses top and bottom for corbels = 41 courses in spiral.

To establish amount of twist per course for the column:

Formula: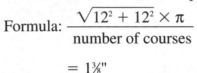

$$\text{Formula: } \frac{\sqrt{12^2 + 12^2} \times \pi}{\text{number of courses}}$$

$$= 1\tfrac{3}{8}"$$

**Figure 31.3**

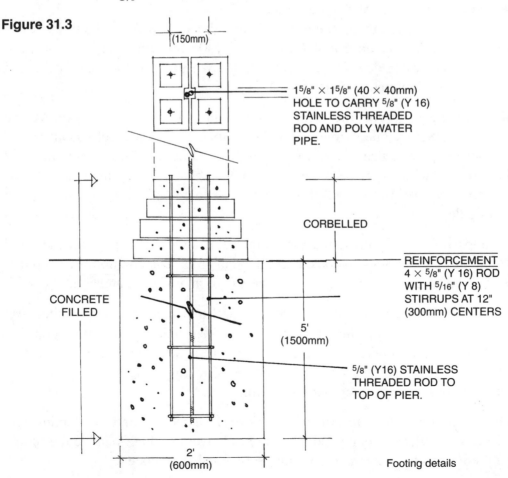

(150mm)

1⁵/₈" × 1⁵/₈" (40 × 40mm) HOLE TO CARRY ⁵/₈" (Y 16) STAINLESS THREADED ROD AND POLY WATER PIPE.

CORBELLED

REINFORCEMENT
4 × ⁵/₈" (Y 16) ROD WITH ⁵/₁₆" (Y 8) STIRRUPS AT 12" (300mm) CENTERS

CONCRETE FILLED

5'
(1500mm)

⁵/₈" (Y16) STAINLESS THREADED ROD TO TOP OF PIER.

2'
(600mm)

Footing details

## Clay Block Spiral Curve (Fig. 31.4B)

To establish number of courses:

Formula: $\dfrac{\text{Radius of curve} \times 2 \times \pi}{4}$   (establish quadrant)

$151.5" - 49.8"$ (3771 − 1240 mm)  (part of quadrant not used)

$\dfrac{101.7"}{3.25"}\left(\dfrac{2531\ \text{mm}}{81\ \text{mm}}\right)$    [(clay brick gauge is 3³⁄₁₆" (81 mm)]

Therefore 31 courses required.

To establish amount of twist per course apply formula above using dimensions of clay block column.

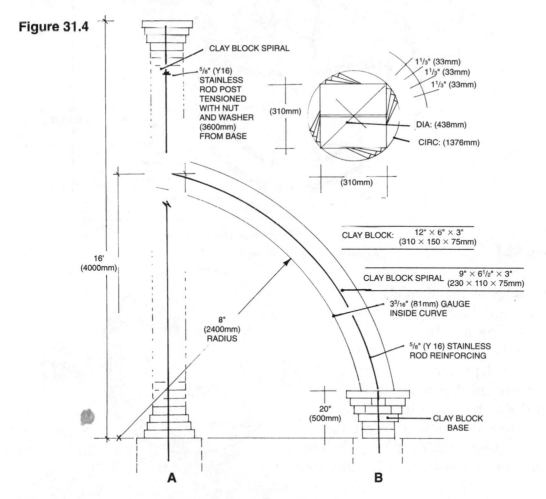

**Figure 31.4**

CLAY BLOCK SPIRAL

⁵⁄₈" (Y16) STAINLESS ROD POST TENSIONED WITH NUT AND WASHER (3600mm) FROM BASE

1¹⁄₃" (33mm)
1¹⁄₃" (33mm)
1¹⁄₃" (33mm)

(310mm)

DIA: (438mm)
CIRC: (1376mm)

(310mm)

CLAY BLOCK:   12" × 6" × 3" (310 × 150 × 75mm)

16' (4000mm)

CLAY BLOCK SPIRAL   9" × 6½" × 3" (230 × 110 × 75mm)

3³⁄₁₆" (81mm) GAUGE INSIDE CURVE

8" (2400mm) RADIUS

⁵⁄₈" (Y 16) STAINLESS ROD REINFORCING

20" (500mm)

CLAY BLOCK BASE

A

B

# 32

# Mandala

## (Spiritual Circle)

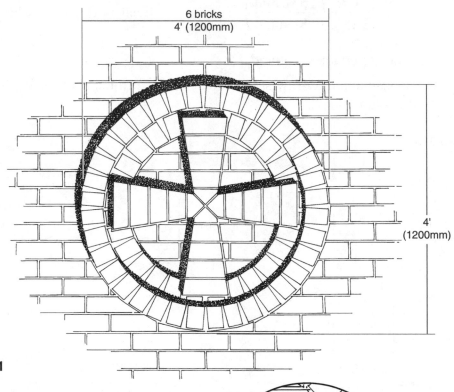

**Figure 32.1**

The circles of the Mandala are formed in the same manner as constructing a bullseye arch. Generally, any size can be obtained. I prefer to lay the brickwork on the inner circle then draw the cross in marker pen, number the bricks to be cut, then brick the inner circle back up the cross which is one full brick to half in the center.

# Part IV
# Brickwork—Tips and Tricks
# 33
# Finishing Off

Raking, brushing, and cleaning down is surely the most important aspect of finishing brickwork. In bricklaying teams, certain men are trained in finishing, and it becomes a trade in itself.

**Figure 33.1**

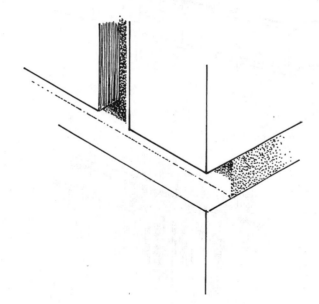

There are certain points I would like to highlight (Fig. 33.1).

1. *Raking of the perpend bed joints:* There should be no hooking of the joints (long furrows of mortar left on the top and bottom of the bed or perpend joint). After raking, stand back and look at the job to see it's been done right.

2. Brush the excess raker marks out.

3. *Cleaning down the wall at the scaffold height is a must*: Windows and doors sprayed previously should preferably be hosed down again on completion of the brickwork. Where the brickwork is to be cleaned by the bricklayer, foam and carpet are ideal. Stopping to rake and clean the brickwork before setting of the mortar prevents scratching of the raked or ironed joints and also the baking of the mortar on the face of the wall, thus making cleaning of the brickwork a simple task.

4. Iron jointing is probably the hardest form of jointing to master. The jointer must be held straight while ironing the bed joints.

**Figure 33.2**

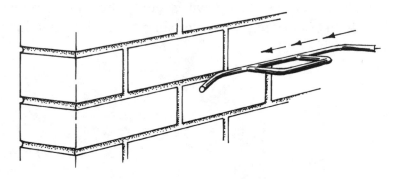

**Figure 33.3**

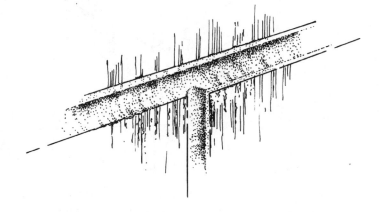

Poor iron jointing can ruin the look of good brick- or blockwork, but ironed joints are still the best method to compact mortar and prevent excessive water problems.

Round ironed jointing is generally specified for clay blockwork as it compresses the mortar giving it better weather protection. As all blockwork should be painted with a sealer on both surfaces, I like to shallow rake the outside face of the wall and use a round joint iron on the inside face of the block. When the joint is raked, care should be taken not to rake too deeply as this can weaken the wall.

Struck jointing (Fig. 33.4) is a good weather joint. Usually, the joint is done with a pointing trowel because it is very seldom we get a nice straight brick. The use of a thin, straight edge to run the pointing trowel along the bed joints is a good help.

**Figure 33.4**

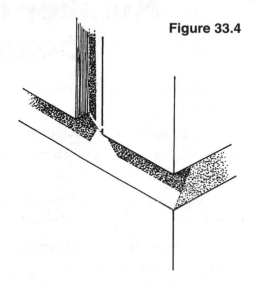

**Figure 33.5**

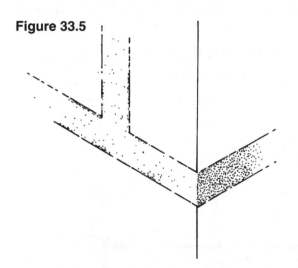

A flush joint (Fig. 33.5) definitely sorts out the riff-raff. Flush joints require good bricklaying. First we need full beds and perps, those brickies who don't double tip the cross joints and don't lay a full bed, generally speaking those who cut corners, will have heaps of trouble. Also good trowel skills in cutting off the mortar without smearing across the face of the brick, good flush and double-sided bricklayers go hand in hand. This brickwork has to be kept clean from the start by the brickies, it is definitely a pain to clean (see Chap. 38).

# 34

# Quoting and Calculating the Number of Bricks in a Construction

Quoting on a job is one of the easiest parts about bricklaying. First, study the plans carefully.

Note the general layout of the construction and ask about things that could inhibit your work: lack of water, lack of power, no vehicle access, buildings on steep sloping ground, etc.

Make sure you take into account all the brickwork. Look for internal walls, fireplaces, internal planter boxes, feature walls with arches, etc.

To determine the number of bricks for a brick-veneer house, first measure the perimeter of the house.

That is, add up the length of all external walls.

**Figure 34.1**

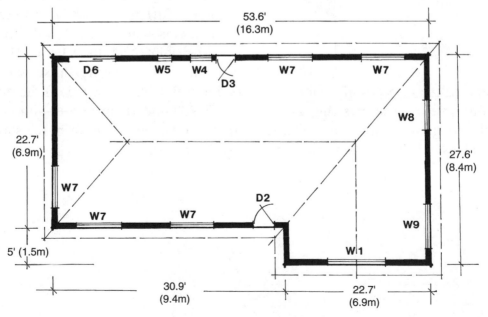

Most houses are 8' (2.4m) high but brickwork starts four courses lower, giving an overall height of 9'1½" (2.735 m).

Multiply the perimeter by the height. This is the total square footage/meters of brickwork. Look for gables to be constructed of brick.

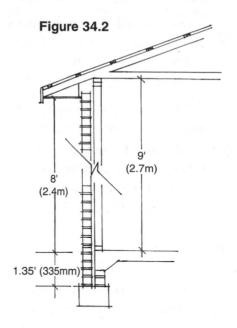

**Figure 34.2**

To work out the square footage/meters in the gable, measure the width by height and divide it by two.

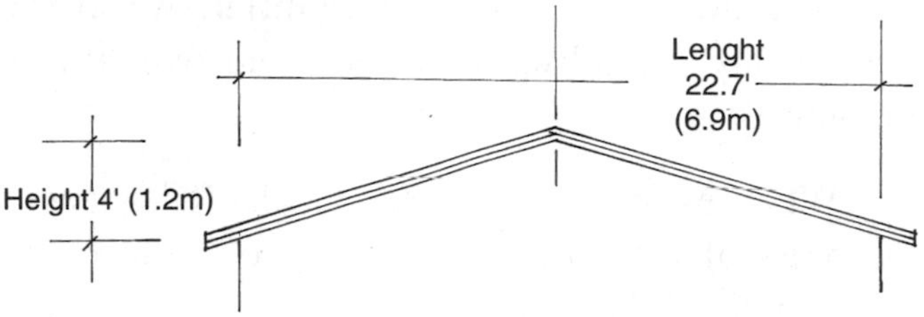

Add up the combined square footage/meters area of all windows and doors by multiplying their length by breadth.

**Figure 34.3**

Subtract this from the total square footage/meters in the construction. Multiplying the area by 63 will give you the total bricks required.

This is based on using good bricks where there will be few rejects. If you know the bricks chosen are likely to have a high rejection rate (because of damage in transit or other reasons), then multiply by 65. This total number of bricks will be sufficient for sills etc.

| W | D | HEIGHT | | LENGTH | |
|---|---|--------|---|--------|---|
| 1 | | 6' (1800mm) | × | 8' | (2400mm) |
| | 2 | 7' (2100mm) | × | 3' | (900mm) |
| | 3 | 7' (2100mm) | × | 2'4" | (700mm) |
| 4 | | 3' (900mm) | × | 3' | (900mm) |
| 5 | | 2' (600mm) | × | 2' | (600mm) |
| | 6 | 7' (2100mm) | × | 6' | (1800mm) |
| 7 | | 4' (1200mm) | × | 6' | (1800mm) |
| 8 | | 4' (1200mm) | × | 4' | (1200mm) |
| 9 | | 6' (1800mm) | × | 6' | (1800mm) |

**Table 34.1**

One further point is to check whether the soffit will be sealed on the rake or on the flat. It can mean the difference between an extra three courses of bricks and going broke. Some home owners ask for different colored bricks along window sills. No problems! Measure the length of each window.

Divide the combined length by 8" (200 mm) and then multiply by three. [There are three bricks to an 8" (200 mm) sill.] In the case of buildings on uneven ground (see following Figures), I use the divided rectangle theory (the same as gables) to work out the requirements. Measure the length by the width and divide by two. You can gauge for yourself whether the slope is greater or less than your halfway decision and accommodate accordingly.

## Materials

About 1.31 cubic yards (1 cubic meter) of loam is needed for 1400 bricks and about 4.5 bags of cement per yard of loam. This calculation is based on average bricks but frog bricks often lay 1500 to the cubic yard. Extruded bricks with exceptionally big holes can come back to as little as 1100 to the cubic yard.

Now to throw down the gauntlet. An average bloke should lay about 500 bricks a day including sills and piers over the duration of the job. (Comments welcome.)

When estimating concrete blocks (8" series) allow 12.5 blocks per yard, also four bags of cement and one bag of lime to one yard of sand. This will lay approximately 600 concrete blocks.

**Figure 34.4**

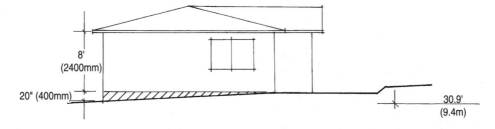

**A** Left Side Elevation

**Figure 34.5**

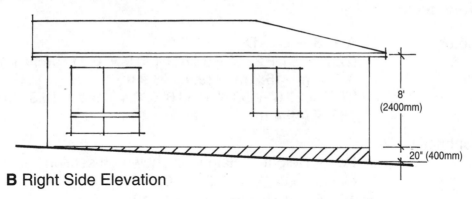

**B** Right Side Elevation

**Figure 34.6**

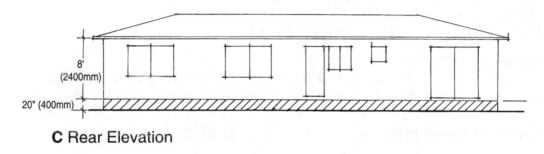

**C** Rear Elevation

**Figure 34.7**

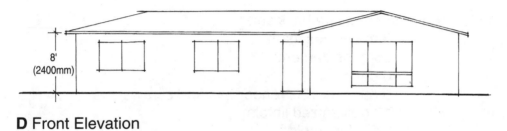

**D** Front Elevation

# Process for determining number of bricks for house shown on previous page

Perimeter
$$= A + B + C + D$$
$$= (22.7' + 5') + 27.6' + 53.6' + (30.9' + 6.9') = (6.9 + 1.5 \text{ m})$$
$$+ 8.4 \text{ m} + 16.3 \text{ m} + (9.4 + 6.9 \text{ m})$$
$$= 27.6' + 27.6' + 53.6' + 53.6' = 8.4 + 8.4 + 16.3 + 16.3 \text{ m}$$
$$= 162.5' \text{ (49.4 m)}$$

Sq ft of brick walls
$$= \text{Per} + \text{Height} - \text{Height as shown on section}$$
elevation 8' + 1.35' (2.4 + 0.344 m)
$$= 162.5' \times 9.35' \text{ (49.4} \times \text{2.744 m)}$$
$$= 1519 \text{ ft}^2 \text{ (135.55 m}^2\text{)}$$

Deduct openings
$$= 1519 - 312.1 \text{ ft}^2 \text{ (135.55} - \text{28.11 m}^2\text{)}$$
$$= 1206.9 \text{ ft}^2 \text{ (107.44 m}^2\text{)}$$

Add brickwork gable
$$= 4' \times 22.7' \div 2 \text{ (1.2 m} \times \text{6.9 m} \div \text{2)}$$
$$= 45.4 \text{ ft}^2 \text{ (4.414 m}^2\text{)}$$

Total sq ft
of brickwork
$$= 1252.3 \text{ ft}^2 \text{ (111.85 m}^2\text{)}$$

Brickwork in footings
$$A = 22.7' \times 1.31' \div 2 = 14.86 \text{ ft}^2 \text{ (6.9} \times \text{.4} \div \text{2} = \text{1.38 m}^2\text{)}$$
$$B = 27.6 \times 1.31 \div 2 = 18 \text{ ft}^2 \text{ (8.4} \times \text{.4} \div \text{2} = \text{1.68 m}^2\text{)}$$
$$C = 53.6 \times 1.31 \div 2 = 35 \text{ ft}^2 \text{ (16.3} \times \text{.4} = \text{6.52 m}^2\text{)}$$

Total of brickwork in footings
$$= 67.86 \text{ft}^2 \text{ (9.58 m}^2\text{)}$$

House and footings
$$1252.3 \text{ ft}^2 \text{ (111.85 m}^2\text{)}$$
$$+ \quad 67.6 \text{ ft}^2 \text{ (9.58 m}^2\text{)}$$

(Total area)
$$= 1319.9 \text{ ft}^2 \div 9\text{ft}^2/\text{yard}^2 = 146.65 \text{ yard}^2 \text{ (121.4 m}^2\text{)}$$
$$= 146.65 \text{ yard}^2 \times 63 \text{ bricks/yard}^2 = 9240 \text{ bricks}$$
$$\text{(121.4 m}^2 \times 50 \text{ bricks/m}^2 = 6070 \text{ bricks)}$$

Materials needed to complete brickwork:
8 m brickies loam
36 bags cement
6 gal color
5 packets 150/veneer ties and screws
13 galvanized lintels
DPC for lintels
11,000 face bricks

# 35
# Contracting for a Living

There is no more true saying than that which goes "I wish I had known then, what I know now." In my case, we're talking about 25 years of brickwork. In days gone by, walls were thicker which meant more "common" bricks were used in between. Commons were quicker and easier to lay and therefore cheaper and more profitable for the bricklayer. Now of course, we lay blocks or bricks straight onto concrete floors.

These days, there is also a lot more excavation which makes life so much easier because there's no more requirement to work in trenches below ground level —but it also reduces profitability because it was this work in the trenches where bricks could be laid quickly, cheaply, and profitably. It is a general trend that in days gone by there were fewer high-set brick houses—now with so many double-story brick houses with double-face brickwork, there is twice as much labor and time required with double-raking of brickwork.

On almost all brick construction (solid brick, cavity brick, clay and concrete block work) the windows are now fitted by the bricklayer, which means more work for which the bricklayer does not get paid. In these changing and competitive times there are also so many rules and requirements. The superannuation levy, payroll tax, work cover, and taxation requirements mean the contract brick layer also has to be a business manager. The taxation system and the ever-changing building codes generally mean you lay bricks in daylight and work on paperwork after hours. I have also discovered that as we get wiser or more experienced we are required to do so much more. Since the earthquakes in Newcastle (Australia), brickies have to fit more ties per square meter. For extra insurance we go overboard with mortaring in damp course. And on most jobs we have to supply our own scaffolding which used to have been supplied by the builder. And how many times have we "helped out the electrician" by fitting some wiring or "run a few pipes" for the plumber? I can even remember a time when the builder would set up the profiles and mark it all down for us. These days of course, builders "on site" are as rare as hen's teeth and we're left to organize most everything, even our own materials. And how much more competitive has the construction industry become? Sometimes the difference between winning and losing a job can be as little as a bag of cement. This is where reputation plays a big part and quality of workmanship is important in succeeding in a competitive field.

I firmly believe that what you miss on the swings you often pick up on the roundabouts so don't be too disheartened if some jobs are not as profitable as they should be. It's better to do the right thing and get repeat business then to cut corners and lose all business.

Technology, however, has helped a lot, especially with plans, ordering equipment, and communication. One of the most important timesavers I have found is the common profile. I can't imagine operating without them but more about this later.

Efficiency is the big factor these days and this comes with planning. Having the site loaded and the gear set up before the job starts and the workers turn up is a must. I operate with four sets of complete gear and I have it all working for me, setting up three jobs ahead and having another set ready for bases.

Little things mean a lot in successful contracting (Fig. 35.1). When loading bases, use stacks of bricks about 1 yard long, then leave a space of about 2'4" (700 mm) for a mortarboard. Say, for instance, the brick base is to be three courses high, load 1-yard-long stacks, two high on their edge. For a four-course base, load 2.5 high, for five courses load 3 high.

And don't waste time by overloading and putting them back. Most importantly, the stacks must be 2' (600 mm) away from the wall of the construction because if they are too close, they will only be in the road and if they are too far away, it will mean too much handling

**Figure 35.1**

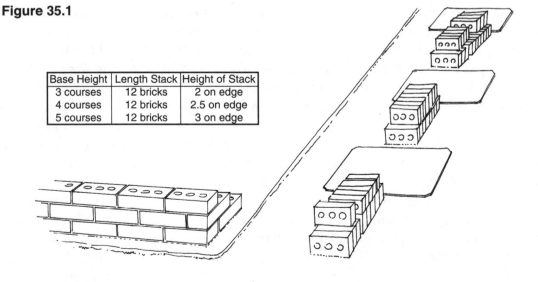

| Base Height | Length Stack | Height of Stack |
|---|---|---|
| 3 courses | 12 bricks | 2 on edge |
| 4 courses | 12 bricks | 2.5 on edge |
| 5 courses | 12 bricks | 3 on edge |

**Figure 35.2**

All planks can be lifted on with a sling, all heavy gear to mixers, cement etc.

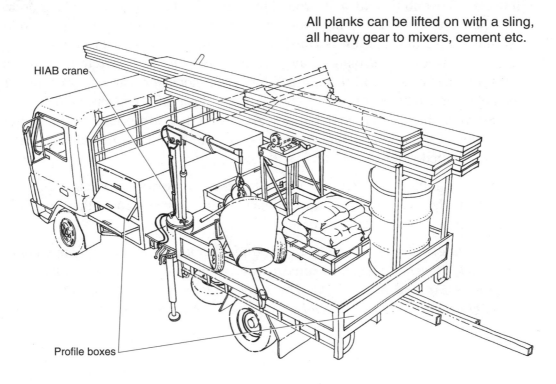

HIAB crane

Profile boxes

Time is money and in our industry we live out of the back of our trucks or at least we seem to. So we have to be on the ball and, starting at the bottom, it's a must to have the right carrying capacity, the right tools, and the right equipment on hand. My own work truck is fitted with a finger crane, which enables my entire crew to keep working up until the last brick is laid while the skilled laborer can load the truck solo. Imagine for a moment what this saves in wages. On a run-of-the-mill domestic lowset house using a team of six bricklayers and three laborers, I have worked out it saves 40 minutes on each job. Multiple that by eight workers at approximately $22.00 per hour and for me it's a saving of $132.00 per job. Over twelve months this is a significant saving. It also has other benefits—the finger crane eliminates all heavy lifting, saving considerably on worker's compensation claims.

I also carry enough scaffold to be able to scaffold half of an average size house—about 40 aluminum planks. When I brick up a high-set house I use an extra 12 planks.

## Part IV: Bricklaying—Tips and Tricks

One small point often overlooked by contractors is finishing the job while on site. It takes almost no time to walk around the job and check for holes in the mortar, to check there are no splits missing under the meter box or the air conditioner, and to be absolutely sure that the job is finished. I find there is absolutely no sense in jumping in the truck and driving away in a hurry only to have to return 20 miles back to a job site to lay two bricks or to finish off a house or patch up a couple of gaps in the mortar. Wherever possible always try to finish off brickwork at the end of the day. So many brickies are tempted to leave off the last two bricks on the corner at scaffold height. It makes it so much harder to set up the line the next day after the two bricks are fitted when the mortar is still wet.

Where toothing is required, lay the extra brick on the corner because this will work wonders when you start next time. Finish off properly and going the extra mile makes the next day's work so much easier.

*Bricklaying is a very physically active trade and here is a good tip for keeping a healthy back. Develop a regular exercise program of sit-ups. It takes no time to do 50 in the morning and 50 at night. The mechanics of this is the muscles in the abdomen help support the back when bending and a strong back will last a lifetime.*

# 36

# Employing Staff

If you are going to run your own business, eventually you will need workers, and I have employed people for over 20 years. From experience, I can usually tell how someone will perform by the way they walk onto the job site. In our game performance counts and not long ago, in fact less than about 15 years ago, you couldn't say to a bricklayer "I'll see you tomorrow," because they probably wouldn't turn up for about 3 days. After a few harsh words the employee would probably say, "Boss, tomorrow never comes. I'm only turning up today to help you out." The end result is that I would have to shake them by the hand and thank them for turning up. Thank god those days are gone. People selection is never easy, it takes practice and patience. I have a small selection test I use. I set up three jobs, two easy ones and a difficult one. If the new person takes the hard one straight up I'll employ him or her straightaway because it's a fair indication that person knows his or her trade, and is not scared of a challenge, and is prepared to get on with the job.

Selection and training of laborers is also important because regardless of how fast or neat a tradesperson is, if the laborer can't keep up the mortar and bricks, then time is wasted. It is also inefficient for bricklayers to be doing the laboring (and they are generally not good at it) so ensuring laborers know their job is vital. One of my laborers does all the cut work, another handles the scaffold and raking, and another does mortar. They have each become good at their individual jobs and we derive great efficiencies.

# 37

# Eliminating Waste of Time

We all know every job has its hard parts, fiddly cut-ins, difficult corners and other aspects, which are generally slow working.

To ensure these jobs are done efficiently, and indeed correctly, you should make sure they are planned well in advance.

The boss on the job should get hold of the plans, know what has to be done, and be able to put the right people to work on the right job.

Often there will be jobs that may be faster to do yourself than to teach someone else to do or to spend time explaining what you want done. I know this is difficult for some brickie bosses, but there are going to be occasions when you're better off getting in and doing it than trying to teach someone else.

The other consideration is that people working slightly outside their capabilities on difficult jobs can cause you a lot of difficulties if their part doesn't come up or tie in with all the rest.

It is much more economical to have one person looking at the plan than six, and as the job progresses you become more "plan wise." If it can't be explained easily, do it yourself.

Pick the people who can do the jobs quickly and match them with compatible work.

After completion of the job leave the site clean, stack up unused bricks, and generally clean up brick rubble.

Keep your tools clean—no bricklayer will regret the small amount of time taken in cleaning tools at the end of a day's work.

Bricklaying is about experience. But it's also about looking, listening, and learning every day on every job. It's not hard to be a bricklayer but it requires a great deal more commitment to be a very good bricklayer and it requires even more to be a skilled, professional tradesperson. Knowledge only comes from doing, making mistakes, and learning from other jobs. I recommend all persons who embark on a bricklaying trade study as many jobs as they can. Very few jobs are ever the same. Know the tricks of the trade, ask questions, read the plans, plan your work, and work your plan.

# 38

# Cleaning Brick- and Blockwork

Our brickwork is kept clean by the laborer with buckets of clean water and sponge or piece of carpet to wipe the face of the brickwork. Timing is essential—stains left too long on a wall in the sun, will bake into the surface. Use of clean water and a sponge is chemical free, simple, and won't cause any further damage.

## Timber Stains

Most timber stains can be removed by using BC 36. I dilute this down to four parts BC to one part water. This mixture is very safe and, if spilled on the concrete, won't cause any harm. At this concentration scrubbing is not necessary. I use a painter's roller tray and a broom, paint it on and leave it for 10 minutes or so. If it needs doing again, you will know. Keep coating until the stains disappear.

A wall with flush joints is difficult to clean without smearing the brick faces. Pieces of foam rubber are used to neatly clean the bricks without crossing the mortar. The same method is used on iron-jointed brickwork or any brickwork where the mortar is flush with the face work. Adding a small amount of acid to the wash will help.

Some brick colors may not be compatible with some colored mortar. White bricks with red or black mortar is an example of a nice contrast job that is difficult to keep clean. High-contrast mortar is a real worry when it splashes onto faces off the scaffold and the newer high-strength mortars are so hard to remove—even with high- pressure cleaners.

## Mortar Stains

Some bricks were not meant to be laid in wet weather. Water running through the top of the wall and passing to the face of the brick can cause smearing. To clean off smearing or mortar stains overlooked by the laborer it is a must to saturate the wall with water until the wall "bubbles." Always start at the bottom of the wall.

Starting at the bottom is important because stains are hard to see on a wet wall. Work small areas 3 to 4 square yards. The saturation process should also stop the acid being drawn into the work. Acid can react with other properties in the clay, and plenty of water in the brick keeps the acid on the surface working at the stains. Wetting the brick prior to aciding will not weaken the acid either. On saturated masonry begin with a mixture of one acid to 20 water and never exceed 1:10.

Leave for 5 minutes for the chemical reaction to break down the stains. Try using a wire or scrubbing brush, making sure to keep plenty of water on the brickwork as it is easy to turn the face black from too much scrubbing. Make sure to hose the acid off the face wall.

All other stains have to be taken off before using hydrochloric acid as vanadium will darken if hydrochloric is used on it, also too much hydrochloric and too little water on dry brickwork can cause burning of the brick work. (See "Acid Burn".)

Due to the wetting of the masonry you may have to leave the wall to dry out for 24 hours to see the results. The above procedure might have to be repeated. Hydrochloric acid is very corrosive and toxic so a protective mask and gloves are a must. I also recommend protective clothing.

*I always maintain that work should be fun but I won't tolerate horseplay on the job. When handling chemicals this is especially important. How many times have we seen apprentices throwing water over each other on a hot day? Let's hope they don't pick up a bucket of acid water by mistake.*

# Vanadium Stain

Vanadium salts occur in the raw materials used in the making of clay products. The yellow and green stains are usually vandyl salts, consisting of sulphates and chlorides, or hydrates of these salts. How these stains appear on the face is as follows. As the water travels through the brick, it dissolves both the vanadium oxides and sulphates. In this process the solution may become very acidic. As it comes through to the surface of the product the salts are deposited.

To remove vanadium I used Noskum—never use hydrochloric acid as this darkens the stain.

# Acid Burn

Acid burns are usually caused by incorrect use of hydrochloric acid on clay masonry. These burns are usually light to deep brown in color, similar to rust. The stains are caused by the hydrochloric acid which penetrates the brick and reacts with the iron oxides in the clay, the iron oxides can also be present in the clay in the loam/mortar mix.

To remove such stains, I use phosphoric acid. It's as dear as poison and twice as effective. This chemical can also destroy the window frame coatings and facia and gutter paint, so use it with care. When the solution is applied mixed in the correct proportions, you don't have to neutralize it, but be careful not to paint any of the solution on the mortar as it will discolor it. Use a paint brush and paint only the affected area. Do not roll the solution as this has the tendency to flick, causing damage.

# 39
# Experiments

I work a lot with our local university testing and experimenting on various construction techniques and it's often interesting just what can be found out when a qualified practitioner and an academic combine their knowledge. I have conducted experiments on the sheer factor of DPC. By placing mortar under the DPC and on top of the DPC (Fig. 39.1), basically sandwiching the DPC between mortar, I find it prevents considerable movement.

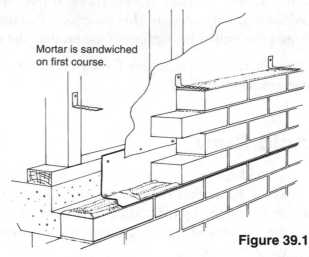

Mortar is sandwiched on first course.

**Figure 39.1**

When pressure is exerted on the wall from either side, say in a cyclone or earthquake, I have found that once the bond is broken the brickwork will move quite easily (Fig. 39.2). This is the accepted way of fitting DPC in Australia to construction standard AS3700. I feel however that by placing ties on top of the first course of bricks as shown in Fig. 39.3 on the DPC, best results are achieved by having a bed of mortar first, then placing the ties on top of the mortar, sandwiching them. And where possible it's even better to bend the end of the tie into the extruded

**Figure 39.2**

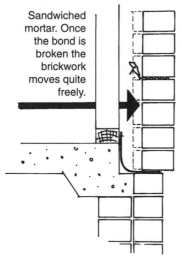

Sandwiched mortar. Once the bond is broken the brickwork moves quite freely.

**Figure 39.3**

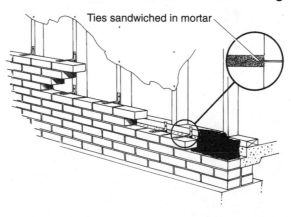

Ties sandwiched in mortar

hole. I found that when pressure was applied on brick walls with standard insertion of ties, the brickwork would push or pull the timber frame under external force. My recommendation is that the ties be placed on top of the first and the third course and thereafter every four courses (Fig. 39.4). There is also an enormous difference with fixing the ties. When pressure is applied to the

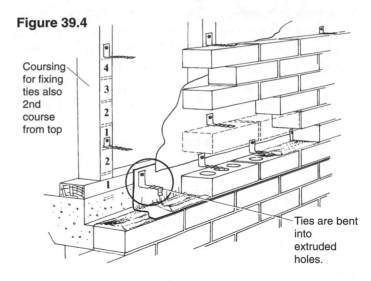

**Figure 39.4**

Coursing for fixing ties also 2nd course from top

Ties are bent into extruded holes.

inside of the wall by whatever means (Figs. 39.5 and 39.6) (in our experiments we use a small hydraulic porta pack), ties that are nailed pull out of the pine framework when they are sandwiched in mortar. This means the sandwiched ties are stronger than the nailed portion. I find that when the ties are placed directly on the brick it creates a slip joint and when pressure is applied the ties stay on the frame although they are ripped out of the brickwork. If anything comes out of all this, it is that all ties, especially in

construction in high wind areas, should be screwed on. And as it is possible for brickwork to flex in high winds I recommend ties should be placed on the last two courses or at least on top of the third course down from the top of the wall. It is possible to pass the ties through the building paper and fix them to the sides of the studs but this can cause water problems.

Talking about reducing the slip joint, these joints can occur at first floor height. The system can also be used anywhere there is a DPC.

The W40 wind category standard for construction is for a pressure loading of up to 1.2 KN.

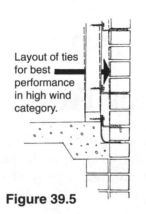

Layout of ties for best performance in high wind category.

**Figure 39.5**

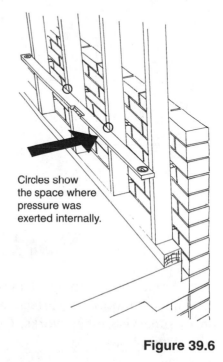

Circles show the space where pressure was exerted internally.

**Figure 39.6**

Most people however fail to realize that in a cyclone this will usually be a .7 force in one direction (pressure) and an opposite reaction of .5 suction on the other side. Another point to remember is that bond strength is not mortar strength. Australian standard AS3700 states that M3 mortar, which is a three to one mix, does not have as much bond strength as an M1 mix. Personally I think because 95 percent of our bricks are now extruded bricks with three huge holes which fill with mortar, we could be looking at a shear strength on bed joints as opposed to bond strength. Therefore in my opinion, M3 mortar is far stronger for shear factor than M1 mortar.

Finding the ultimate strength in mortar comes back to the ingredients. The strongest mortar is that made with washed sand. Any clay particles in the sand will weaken the mortar. Clay does however appear to make mortar seem more "fluid" or "plasticey" but the downside is the loss of compressive and adhesive strength. There has been a trend lately to adding fire clay to mortar mixes but I don't recommend it.

**Table 39.1**

| Mortar Classifications | | | |
|---|---|---|---|
| **Classification** | **Type A Portland Cement** | **Building Line** | **Sand** |
| M1 | 0 | 1 | 3 |
| M2 | 1 | 2 | 9 |
|  | 1 | 3 | 12 |
| M3 | 1 | 1 | 6 |
|  | 1 | 0 + water thickener | 5 |
| M4 | 1 | 0 to 0.25 | 3 |
|  | 1 | 0.5 | 4.5 |
|  | 1 | 0 + water thickener | 4 |

# 40

# Further Reading

Frank Gildbreth was a master tradesman bricklayer. He also invented studies on time and motion. So effective was he at his trade that many companies kept him on to supervise other works. Get yourself a copy of "*Writings of Gildbreth*"—it will inspire you.

# 41

# Calculations

**Figure 41.1**

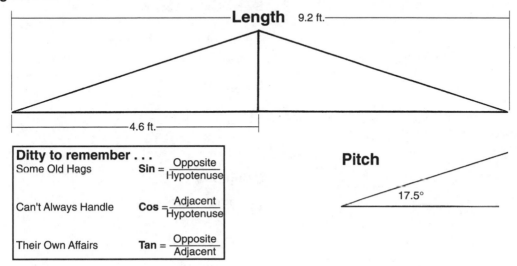

**Length** 9.2 ft.

4.6 ft.

| Ditty to remember . . . | | |
|---|---|---|
| Some Old Hags | **Sin** = | $\dfrac{\text{Opposite}}{\text{Hypotenuse}}$ |
| Can't Always Handle | **Cos** = | $\dfrac{\text{Adjacent}}{\text{Hypotenuse}}$ |
| Their Own Affairs | **Tan** = | $\dfrac{\text{Opposite}}{\text{Adjacent}}$ |

**Pitch**

17.5°

**Figure 41.2**

**To find:**

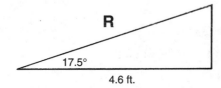

R

17.5°

4.6 ft.

$$R = \frac{4.6 \text{ ft.}}{\cos 17.5°} = 4.823 \text{ ft.}$$

**Figure 41.3**

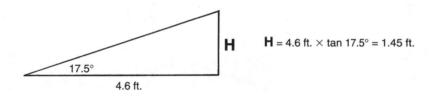

H

17.5°

4.6 ft.

$$H = 4.6 \text{ ft.} \times \tan 17.5° = 1.45 \text{ ft.}$$

## To Find:

### Figure 41.4

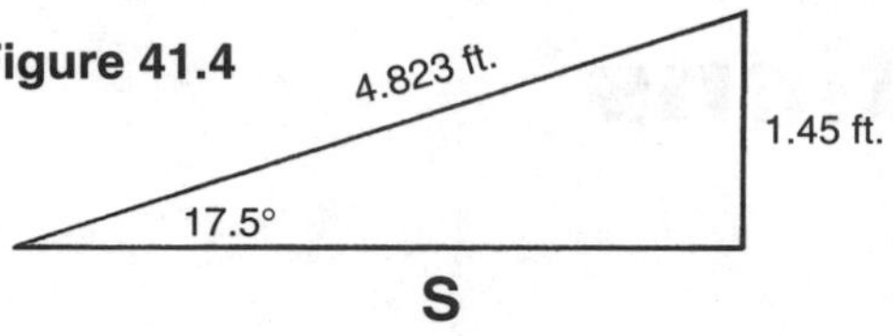

4.823 ft.

1.45 ft.

17.5°

S

### Using Tan

$$S = \frac{\text{opposite side}}{\text{tan of angle}}$$

$$S = \frac{1.45 \text{ ft.}}{\text{tan } 17.5°}$$

### Using Cos

$$S = \text{hypotenuse} \times \text{cos of angle}$$

$$S = 4.823 \text{ ft.} \times \cos 17.5° = 4.6 \text{ ft.}$$

### Figure 41.5

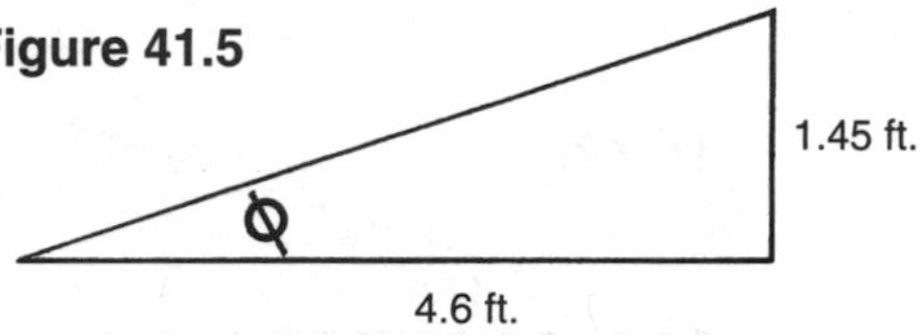

1.45 ft.

Q

4.6 ft.

### Using Tan

$$Q = \frac{\text{opposite side}}{\text{adjacent tan}}$$

$$Q = \frac{1.45 \text{ ft.}}{4.6 \text{ ft.}} = 17.5$$

### Using Cos

$$Q = \frac{\text{adjacent side}}{\text{hypotenuse}} = \cos$$

$$Q = \frac{4.6 \text{ ft.}}{4.823 \text{ ft.}} = 17.5$$

## To find length of any side of a triangle:

### Figure 41.6

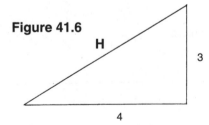

H

3

4

$$H = \sqrt{L^2 + B^2}$$
$$= \sqrt{4 \times 4 + 3 \times 3} = \sqrt{25} = 5$$

### Figure 41.7

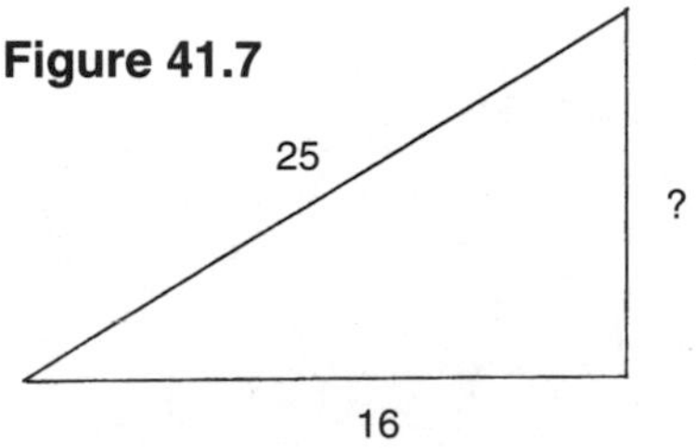

25

?

16

$$? = \sqrt{25 - 16} = \sqrt{9} = 3$$

## Quick Calculations

• To multiply anything by 5, just halve it then add a zero.

• To divide anything by 5, just divide it by 10 then double it.

• To multiply by 11, add the 2 digits together and place the answer in the middle. (e.g., 72 × 11 = 7 + 2 = 9 therefore 792.) If it comes to 10 or more, add to the left digit. (e.g., 86 × 11 = 8 + 6 = 14 therefore 946.)

• Any numbers that add up to 9 are evenly divisible by 3. (e.g., 162 = 1 + 6 + 2 = 9 therefore 162 divided by 3 = 54.)

**Figure 41.8**

TRIANGLE

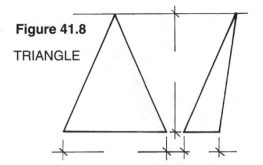

$$AREA = \frac{B \times Ht}{2}$$

**Figure 41.9**

TRAPEZOID

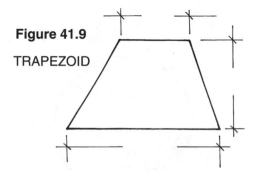

$$AREA = \frac{(A+B) \times Ht}{2}$$

**Figure 41.10**

CIRCLE

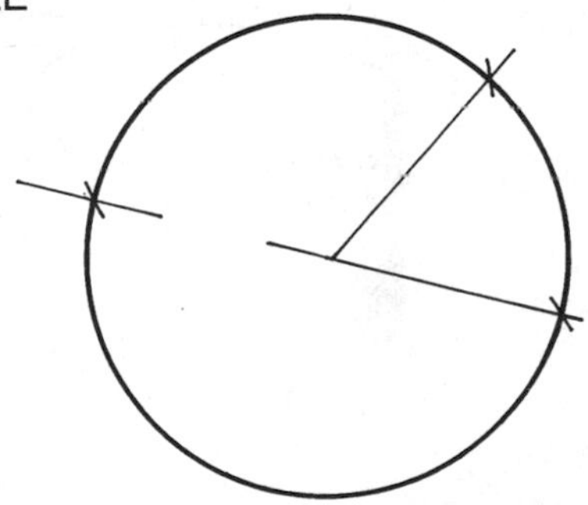

$$AREA = 3.1416 \times R^2$$

**Figure 41.11**

RECTANULAR PRISM

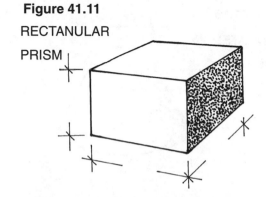

$$VOLUME = A \times B \times C$$

**Figure 41.12**

TRIANGULAR
PYRAMID

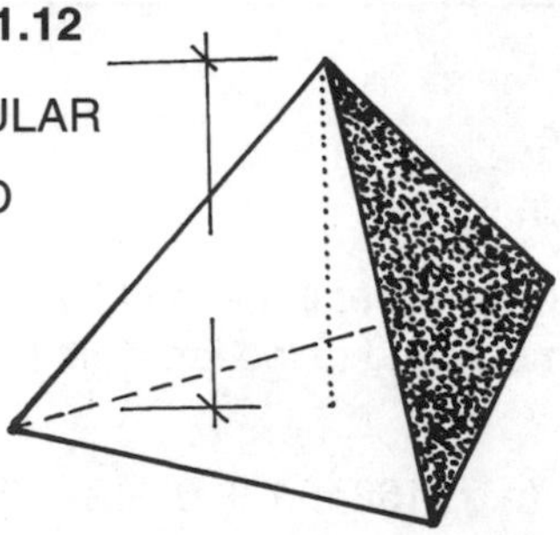

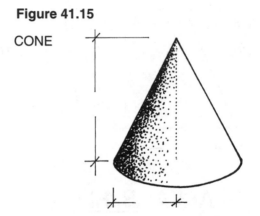

$$\text{VOLUME} = \frac{\text{Area of Base x Ht}}{3}$$

**Figure 41.13**

TRUNCATED
PYRAMID

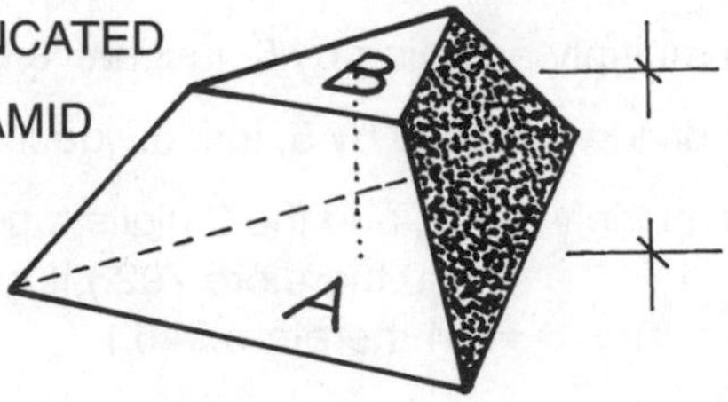

$$\text{VOLUME} = \frac{\text{Ht x }(A+B+\sqrt{A \times B})}{3}$$

**Figure 41.14**

SPHERE

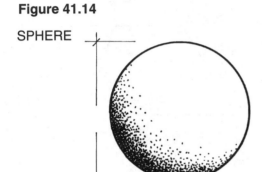

$$\text{VOLUME} = \frac{4 \times 3.1416 \times R^3}{3}$$

**Figure 41.15**

CONE

$$\text{VOLUME} = \frac{3.1416 \times R^2 \times \text{Ht}}{3}$$

**Figure 41.16**

TRUNCATED
CONE

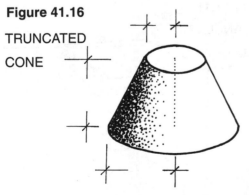

$$\text{VOLUME} = \frac{3.1416 \times \text{Ht} \times (R^2 + r^2 + [R \times r])}{3}$$

**Figure 41.17**

CYLINDER

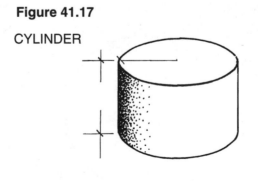

$$\text{VOLUME} = 3.1416 \times R^2 \times \text{Ht}$$

**Figure 41.18**

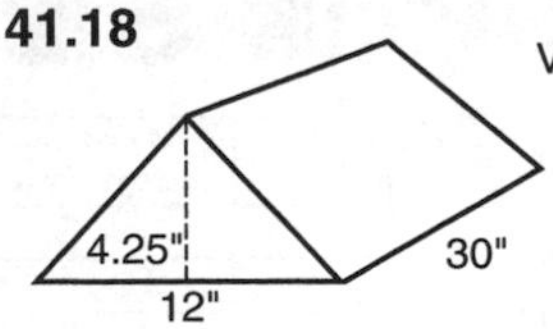

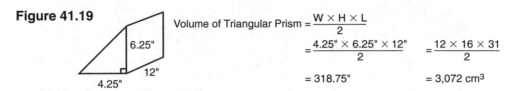

Volume of Triangular Prism $= \dfrac{W}{2} \times H \times L$

$= 6" \times 4.25" \times 30"$    $= 15.5 \times 11 \times 7$

$= 765"$    $= 12,958 \text{ cm}^3$

**Figure 41.19**

Volume of Triangular Prism $= \dfrac{W \times H \times L}{2}$

$= \dfrac{4.25" \times 6.25" \times 12"}{2}$    $= \dfrac{12 \times 16 \times 31}{2}$

$= 318.75"$    $= 3,072 \text{ cm}^3$

# Using Tables for Calculation

Refer to tables corresponding to length required and brick size for number of bricks required. Lengths shown for number of bricks are based on standard mortar width of ⅜". Should scale length shown be less than required length, divide difference by number of bricks—1 and add the result to standard mortar width. Similarly, if scale length shown is more than required length, divide difference by number of bricks—1, and subtract result from standard mortar width.

## Example:

Wall length   33'

Using 200/8" bricks nearest scale length = 32' 11"

For 49 bricks — divide 3" scale length by 49 = ¹⁄₁₆"; so increase mortar width to ⁷⁄₁₆".

If formula use preferred:

Let  $x$ = required length
     $y$ = scale length closest to required length
     $a$ = number of bricks

Where required length exceeds scale length:

$$\frac{x - y}{a - 1}$$

Where scale length exceeds required length:

$$\frac{y - x}{a - 1}$$

Should difference be 3" to 4½" (73 to 115 mm) either way, add or subtract half a brick.

# Appendix A. Table of Brick Sizes

| Course Length | Feet and Inches Increments 7¾ | | | Feet and Inches Increments 7³/₁₆ | | | Feet and Inches Increments 7⅞ | | | Feet and Inches Increments 7⁵/₁₆ | | | Feet and Inches Increments 8 | | | Feet and Inches Increments 8¹/₁₆ | | |
|---|---|---|---|---|---|---|---|---|---|---|---|---|---|---|---|---|---|---|
| 1 | | 7 | ¾ | | 7 | ³/₁₆ | | 7 | ⅞ | | 7 | ⁵/₁₆ | | 8 | | | 8 | ¹/₁₆ |
| 2 | 1 | 3 | ½ | 1 | 3 | ⁵/₈ | 1 | 3 | ¾ | 1 | 3 | ⅞ | 1 | 4 | | 1 | 4 | ¹/₈ |
| 3 | 1 | 11 | ¼ | 1 | 11 | ⁷/₁₆ | 1 | 11 | ⅝ | 1 | 11 | ¹³/₁₆ | 2 | 0 | | 2 | 0 | ³/₁₆ |
| 4 | 2 | 7 | | 2 | 7 | ¼ | 2 | 7 | ½ | 2 | 7 | ¾ | 2 | 8 | | 2 | 8 | ¼ |
| 5 | 3 | 2 | ¾ | 3 | 3 | ¹/₁₆ | 3 | 3 | ⅜ | 3 | 3 | ¹¹/₁₆ | 3 | 4 | | 3 | 4 | ⁵/₁₆ |
| 6 | 3 | 10 | ½ | 3 | 10 | ⁷/₈ | 3 | 11 | ¼ | 3 | 11 | ⁵/₈ | 4 | 0 | | 4 | 0 | ³/₈ |
| 7 | 4 | 6 | ¼ | 4 | 6 | ¹¹/₁₆ | 4 | 7 | ⅛ | 4 | 7 | ⁹/₁₆ | 4 | 8 | | 4 | 8 | ⁷/₁₆ |
| 8 | 5 | 2 | | 5 | 2 | ½ | 5 | 3 | | 5 | 3 | ½ | 5 | 4 | | 5 | 4 | ¼ |
| 9 | 5 | 9 | ¾ | 5 | 10 | ⁵/₁₆ | 5 | 10 | ⅞ | 5 | 11 | ⁷/₁₆ | 6 | 0 | | 6 | 0 | ⁹/₁₆ |
| 10 | 6 | 5 | ½ | 6 | 6 | ¹/₈ | 6 | 6 | ¾ | 6 | 7 | ³/₈ | 6 | 8 | | 6 | 8 | ⁵/₈ |
| 11 | 7 | 1 | ¼ | 7 | 1 | ¹⁵/₁₆ | 7 | 2 | ⅝ | 7 | 2 | ⁵/₁₆ | 7 | 4 | | 7 | 4 | ¹¹/₁₆ |
| 12 | 7 | 9 | | 7 | 9 | ¾ | 7 | 10 | ½ | 7 | 11 | ¼ | 8 | 0 | | 8 | 0 | ¾ |
| 13 | 8 | 4 | ¾ | 8 | 5 | ⁹/₁₆ | 8 | 6 | ⅜ | 8 | 7 | ³/₁₆ | 8 | 8 | | 8 | 8 | ¹³/₁₆ |
| 14 | 9 | 0 | ½ | 9 | 1 | ³/₈ | 9 | 2 | ¼ | 9 | 3 | ⅛ | 9 | 4 | | 9 | 4 | ⁷/₈ |
| 15 | 9 | 8 | ¼ | 9 | 9 | ³/₁₆ | 9 | 10 | ⅛ | 9 | 11 | ¹/₁₆ | 10 | 0 | | 10 | 0 | ¹⁵/₁₆ |
| 16 | 10 | 4 | | 10 | 5 | | 10 | 6 | | 10 | 7 | | 10 | 8 | | 10 | 9 | |
| 17 | 10 | 11 | ¾ | 11 | 0 | ¹³/₁₆ | 11 | 1 | ⅞ | 11 | 2 | ¹⁵/₁₆ | 11 | 4 | | 11 | 5 | ¹/₁₆ |
| 18 | 11 | 7 | ½ | 11 | 8 | ⁵/₈ | 11 | 9 | ¾ | 11 | 10 | ⅞ | 12 | 0 | | 12 | 1 | ¹/₈ |
| 19 | 12 | 3 | ¼ | 12 | 4 | ⁷/₁₂ | 12 | 5 | ⅝ | 12 | 6 | ¹³/₁₆ | 12 | 8 | | 12 | 9 | ³/₁₆ |
| 20 | 12 | 11 | | 13 | 0 | ¼ | 13 | 1 | ½ | 13 | 2 | ¾ | 13 | 4 | | 13 | 5 | ¼ |
| 21 | 13 | 6 | ¾ | 13 | 8 | ¹/₁₆ | 13 | 9 | ⅜ | 13 | 10 | ¹¹/₁₆ | 14 | 0 | | 14 | 1 | ⁵/₁₆ |
| 22 | 14 | 2 | ½ | 14 | 3 | ⁷/₈ | 14 | 5 | ¼ | 14 | 6 | ⅝ | 14 | 8 | | 14 | 9 | ³/₈ |
| 23 | 14 | 10 | ¼ | 14 | 11 | ¹¹/₁₆ | 15 | 1 | ⅛ | 15 | 2 | ⁹/₁₆ | 15 | 4 | | 15 | 5 | ⁷/₁₆ |
| 24 | 15 | 6 | | 15 | 7 | ½ | 15 | 9 | | 15 | 10 | ½ | 16 | 0 | | 16 | 1 | ½ |
| 25 | 16 | 1 | ¾ | 16 | 3 | ⁵/₁₆ | 16 | 4 | ⅞ | 16 | 6 | ⁷/₁₆ | 16 | 8 | | 16 | 9 | ⁹/₁₆ |
| 26 | 16 | 11 | ½ | 16 | 11 | ¹/₈ | 17 | 0 | ¾ | 17 | 2 | ³/₈ | 17 | 4 | | 17 | 5 | ⁵/₈ |
| 27 | 17 | 5 | ¼ | 17 | 6 | ¹⁵/₁₆ | 17 | 8 | ⅝ | 17 | 10 | ⁵/₁₆ | 18 | 0 | | 18 | 1 | ¹¹/₁₆ |
| 28 | 18 | 1 | | 18 | 2 | ¾ | 18 | 4 | ½ | 18 | 6 | ¼ | 18 | 8 | | 18 | 9 | ¾ |
| 29 | 18 | 8 | ¾ | 18 | 10 | ⁹/₁₆ | 19 | 0 | ⅜ | 19 | 2 | ³/₁₆ | 19 | 4 | | 19 | 5 | ¹³/₁₆ |
| 30 | 19 | 4 | ½ | 19 | 6 | ³/₈ | 19 | 8 | ¼ | 19 | 10 | ⅛ | 20 | 0 | | 20 | 1 | ⁷/₈ |
| 31 | 20 | 0 | ¼ | 20 | 2 | ³/₁₆ | 20 | 4 | ⅛ | 20 | 6 | ¹/₁₆ | 20 | 8 | | 20 | 9 | ¹⁵/₁₆ |
| 32 | 20 | 8 | | 20 | 10 | | 21 | 0 | | 21 | 2 | | 21 | 4 | | 21 | 6 | |
| 33 | 21 | 3 | ¾ | 21 | 5 | ¹³/₁₆ | 21 | 7 | ⅞ | 21 | 9 | ¹⁵/₁₆ | 22 | 0 | | 22 | 2 | ¹/₁₆ |
| 34 | 21 | 11 | ½ | 22 | 1 | ⁵/₈ | 22 | 3 | ¾ | 22 | 5 | ⅞ | 22 | 8 | | 22 | 10 | ¹/₈ |
| 35 | 22 | 5 | ¼ | 22 | 9 | ⁷/₁₆ | 22 | 11 | ⅝ | 23 | 1 | ¹³/₁₆ | 23 | 4 | | 23 | 6 | ³/₁₆ |
| 36 | 23 | 3 | | 23 | 5 | ¼ | 23 | 7 | ½ | 23 | 9 | ¾ | 24 | 0 | | 24 | 2 | ¼ |
| 37 | 23 | 10 | ¾ | 24 | 1 | ¹/₁₆ | 24 | 3 | ⅜ | 24 | 5 | ¹¹/₁₆ | 24 | 8 | | 24 | 10 | ⁵/₁₆ |
| 38 | 24 | 6 | ½ | 24 | 8 | ⁷/₈ | 24 | 10 | ¼ | 25 | 1 | ⅝ | 25 | 4 | | 25 | 6 | ³/₈ |
| 39 | 25 | 2 | ¼ | 25 | 4 | ¹¹/₁₆ | 25 | 7 | ⅛ | 25 | 9 | ⁹/₁₆ | 26 | 0 | | 26 | 2 | ⁷/₁₆ |
| 40 | 25 | 10 | | 26 | 0 | ½ | 26 | 3 | | 26 | 5 | ½ | 26 | 8 | | 26 | 10 | ½ |
| 41 | 26 | 5 | ¾ | 26 | 8 | ⁵/₁₆ | 26 | 10 | ⅞ | 27 | 1 | ⁷/₁₆ | 27 | 4 | | 27 | 6 | ⁹/₁₆ |
| 42 | 27 | 1 | ½ | 27 | 4 | ¹/₈ | 27 | 6 | ¾ | 27 | 9 | ³/₈ | 28 | 0 | | 28 | 2 | ⁵/₈ |
| 43 | 27 | 9 | ¼ | 27 | 11 | ¹⁵/₁₆ | 28 | 2 | ⅝ | 28 | 5 | ⁵/₁₆ | 28 | 8 | | 28 | 10 | ¹¹/₁₆ |
| 44 | 28 | 5 | | 28 | 7 | ¾ | 28 | 10 | ½ | 29 | 1 | ¼ | 29 | 4 | | 29 | 6 | ¾ |
| 45 | 29 | 0 | ¾ | 29 | 3 | ⁹/₁₆ | 29 | 6 | ⅜ | 29 | 9 | ³/₁₆ | 30 | 0 | | 30 | 2 | ¹³/₁₆ |
| 46 | 29 | 8 | ½ | 29 | 11 | ³/₈ | 30 | 2 | ¼ | 30 | 5 | ⅛ | 30 | 8 | | 30 | 10 | ⁷/₈ |
| 47 | 30 | 4 | ¼ | 30 | 7 | ³/₁₆ | 30 | 10 | ⅛ | 31 | 1 | ¹/₁₆ | 31 | 4 | | 31 | 6 | ¹⁵/₁₆ |
| 48 | 31 | 0 | | 31 | 3 | | 31 | 6 | | 31 | 9 | | 32 | 0 | | 32 | 3 | |
| 49 | 31 | 7 | ¾ | 31 | 10 | ¹³/₁₆ | 32 | 1 | ⅞ | 32 | 4 | ¹⁵/₁₆ | 32 | 8 | | 32 | 11 | ¹/₁₆ |
| 50 | 32 | 3 | ½ | 32 | 6 | ⁵/₈ | 32 | 9 | ¾ | 33 | 0 | ⁷/₈ | 33 | 4 | | 33 | 7 | ¹/₈ |

| Course Length | Feet and Inches Increments 7 ¾ | Feet and Inches Increments 7 3/16 | Feet and Inches Increments 7 7/8 | Feet and Inches Increments 7 5/16 | Feet and Inches Increments 8 | Feet and Inches Increments 8 1/16 |
|---|---|---|---|---|---|---|
| 51 | 32 11 ¼ | 33 2 7/16 | 33 5 5/8 | 33 8 13/16 | 34 0 | 34 3 3/16 |
| 52 | 33 7 | 33 10 ¼ | 34 1 ½ | 34 4 ¾ | 34 8 | 34 11 ¼ |
| 53 | 34 2 ¾ | 34 6 1/16 | 34 9 3/8 | 35 0 11/16 | 35 4 | 35 7 5/16 |
| 54 | 34 10 ½ | 35 1 7/8 | 35 5 ¼ | 35 8 5/8 | 36 0 | 36 3 3/8 |
| 55 | 35 6 ¼ | 35 9 11/16 | 36 1 1/8 | 36 4 9/16 | 36 8 | 36 11 7/16 |
| 56 | 36 2 | 36 5 ½ | 36 9 | 37 0 ½ | 37 4 | 37 7 ½ |
| 57 | 36 9 ¾ | 37 1 5/16 | 37 4 7/8 | 37 8 7/16 | 38 0 | 38 3 9/16 |
| 58 | 37 5 ½ | 37 9 1/8 | 38 0 ¾ | 38 4 3/8 | 38 8 | 38 11 5/8 |
| 59 | 38 1 ¼ | 38 4 15/16 | 38 8 5/8 | 39 0 5/16 | 39 4 | 39 7 11/16 |
| 60 | 38 9 | 39 0 ¾ | 39 4 ½ | 39 8 ¼ | 40 0 | 40 3 ¾ |
| 61 | 39 4 ¾ | 39 8 9/16 | 40 0 3/8 | 40 4 3/16 | 40 8 | 40 11 13/16 |
| 62 | 40 0 ½ | 40 4 3/8 | 40 8 ¼ | 41 0 1/8 | 41 4 | 41 7 7/8 |
| 63 | 40 8 ¼ | 41 0 3/16 | 41 4 1/8 | 41 8 1/16 | 42 0 | 42 2 15/16 |
| 64 | 41 4 | 41 8 | 42 0 | 42 4 | 42 8 | 43 0 |
| 65 | 41 11 ¾ | 42 3 13/16 | 42 7 7/8 | 42 11 15/16 | 43 4 | 43 8 1/16 |
| 66 | 42 7 ½ | 42 11 5/8 | 43 3 ¾ | 43 7 7/8 | 44 0 | 44 4 |
| 67 | 43 3 ¼ | 43 7 7/16 | 43 11 5/8 | 44 3 13/16 | 44 8 | 45 0 3/16 |
| 68 | 43 11 | 44 3 ¼ | 44 7 ½ | 44 11 ¾ | 45 4 | 45 8 ¼ |
| 69 | 44 6 ¾ | 44 11 1/16 | 45 3 3/8 | 45 7 11/16 | 46 0 | 46 4 5/16 |
| 70 | 45 2 ½ | 45 6 7/8 | 45 11 ¼ | 46 3 5/8 | 46 8 | 47 0 3/8 |
| 71 | 45 10 ¼ | 46 2 11/16 | 46 7 1/8 | 46 11 9/16 | 47 4 | 47 8 7/16 |
| 72 | 46 6 | 46 10 ½ | 47 3 | 47 7 ½ | 48 0 | 48 4 ½ |
| 73 | 47 1 ¾ | 47 6 5/16 | 47 10 7/8 | 48 3 7/16 | 48 8 | 49 0 9/16 |
| 74 | 47 9 ½ | 48 2 1/8 | 48 6 ¾ | 48 11 3/8 | 49 4 | 49 8 5/8 |
| 75 | 48 5 ¼ | 48 9 15/16 | 49 2 5/8 | 49 7 5/16 | 50 0 | 50 4 11/16 |
| 76 | 49 1 | 49 5 ¾ | 49 10 ½ | 50 3 ¼ | 50 8 | 51 0 ¾ |
| 77 | 49 8 ¾ | 50 1 9/16 | 50 6 3/8 | 50 11 3/16 | 51 4 | 51 8 13/16 |
| 78 | 50 4 ½ | 50 9 3/8 | 51 2 ¼ | 51 7 1/8 | 52 0 | 52 4 7/8 |
| 79 | 51 0 ¼ | 51 5 3/16 | 51 10 1/8 | 52 3 1/16 | 52 8 | 53 0 15/16 |
| 80 | 51 8 | 52 1 | 52 6 | 52 11 | 53 4 | 53 9 |
| 81 | 52 3 ¾ | 52 8 13/16 | 53 1 7/8 | 53 6 15/16 | 54 0 | 54 5 1/16 |
| 82 | 52 11 ½ | 53 4 5/8 | 53 9 ¾ | 54 2 7/8 | 54 8 | 55 1 1/8 |
| 83 | 53 7 ¼ | 54 0 7/16 | 54 5 5/8 | 54 10 13/16 | 55 4 | 55 9 3/16 |
| 84 | 54 3 | 54 8 ¼ | 55 1 ½ | 55 6 ¾ | 56 0 | 56 5 ¼ |
| 85 | 54 10 ¾ | 55 4 1/16 | 55 9 3/8 | 56 2 11/16 | 56 8 | 57 1 5/16 |
| 86 | 55 6 ½ | 55 11 7/8 | 56 5 ¼ | 56 10 5/8 | 57 4 | 57 9 3/8 |
| 87 | 56 2 ¼ | 56 7 11/16 | 57 1 1/8 | 57 6 9/16 | 58 0 | 58 5 7/16 |
| 88 | 56 10 | 57 3 ½ | 57 9 | 58 2 ½ | 58 8 | 59 1 ½ |
| 89 | 56 5 ¾ | 57 11 5/16 | 58 4 7/8 | 58 10 7/16 | 59 4 | 59 9 9/16 |
| 90 | 58 1 ½ | 58 7 1/8 | 59 0 ¾ | 59 6 3/8 | 60 0 | 60 5 5/8 |
| 91 | 58 9 ¼ | 59 2 15/16 | 59 8 5/8 | 60 2 5/16 | 60 8 | 61 1 11/16 |
| 92 | 59 5 | 59 10 ¾ | 60 4 ¾ | 60 10 ¼ | 61 4 | 61 9 ¾ |

# Appendix B. Brick Gauge

| Course Height | Mm Incr 67mm | Ft and Inches Increments 2 2/3 | Ft and Inches Increments 3 | Ft and Inches Increments 3 1/16 | Ft and Inches Increments 3 1/8 | Ft and Inches Increments 3 3/16 |
|---|---|---|---|---|---|---|
| 1 | 67 | 2 2/3 | 3 | 3 1/16 | 3 1/8 | 3 3/16 |
| 2 | 133 | 5 1/3 | 6 | 6 1/8 | 6 1/4 | 6 3/8 |
| 3 | 200 | 8 | 9 | 9 3/16 | 9 3/8 | 9 9/16 |
| 4 | 267 | 10 2/3 | 1 0 | 1 0 1/4 | 1 0 1/2 | 1 0 3/4 |
| 5 | 333 | 1 1 1/2 | 1 3 | 1 3 5/16 | 1 3 5/8 | 1 3 15/16 |
| 6 | 400 | 1 4 | 1 6 | 1 6 3/8 | 1 6 3/4 | 1 7 1/8 |
| 7 | 467 | 1 6 2/3 | 1 9 | 1 9 7/16 | 1 9 7/8 | 1 10 5/16 |
| 8 | 533 | 1 9 1/2 | 2 0 | 2 0 1/2 | 2 1 | 2 1 1/2 |
| 9 | 600 | 2 0 | 2 3 | 2 3 9/16 | 2 4 1/8 | 2 4 11/16 |
| 10 | 667 | 2 2 2/3 | 2 6 | 2 6 5/8 | 2 7 1/4 | 2 7 7/8 |
| 11 | 733 | 2 5 1/3 | 2 9 | 2 9 11/16 | 2 10 3/8 | 2 11 1/16 |
| 12 | 800 | 2 8 | 3 0 | 3 0 3/4 | 3 1 1/2 | 3 2 1/4 |
| 13 | 867 | 2 10 2/3 | 3 3 | 3 3 13/16 | 3 4 5/8 | 3 5 7/16 |
| 14 | 933 | 3 1 1/3 | 3 6 | 3 6 7/8 | 3 7 3/4 | 3 8 5/8 |
| 15 | 1000 | 3 4 | 3 9 | 3 9 15/16 | 3 10 7/8 | 3 11 13/16 |
| 16 | 1067 | 3 6 2/3 | 4 0 | 4 1 | 4 2 | 4 3 |
| 17 | 1133 | 3 9 1/3 | 4 3 | 4 4 1/16 | 4 5 1/8 | 4 6 3/16 |
| 18 | 1200 | 4 0 | 4 6 | 4 7 1/8 | 4 8 1/4 | 4 9 3/8 |
| 19 | 1267 | 4 2 2/3 | 4 9 | 4 10 3/16 | 4 11 3/8 | 5 0 9/16 |
| 20 | 1333 | 4 5 1/3 | 5 0 | 5 1 1/4 | 5 2 1/2 | 5 3 3/4 |
| 21 | 1400 | 4 8 | 5 3 | 5 4 5/16 | 5 5 5/8 | 5 6 15/16 |
| 22 | 1467 | 4 10 2/3 | 5 6 | 5 7 3/8 | 5 8 3/4 | 5 10 1/8 |
| 23 | 1533 | 5 1 1/3 | 5 9 | 5 10 7/16 | 5 11 7/8 | 6 1 5/16 |
| 24 | 1600 | 5 4 | 6 0 | 6 1 1/2 | 6 3 | 6 4 1/2 |
| 25 | 1667 | 5 6 2/3 | 6 3 | 6 4 9/16 | 6 6 1/8 | 6 7 11/16 |
| 26 | 1733 | 5 9 1/3 | 6 6 | 6 7 5/8 | 6 9 1/4 | 6 10 7/8 |
| 27 | 1800 | 6 0 | 6 9 | 6 10 11/16 | 7 0 3/8 | 7 2 1/16 |
| 28 | 1867 | 6 2 2/3 | 7 0 | 7 1 3/4 | 7 3 1/2 | 7 5 1/4 |
| 29 | 1933 | 6 5 1/3 | 7 3 | 7 4 13/16 | 7 6 5/8 | 7 8 7/16 |
| 30 | 2000 | 6 8 | 7 6 | 7 7 7/8 | 7 9 3/4 | 7 11 5/8 |
| 31 | 2067 | 6 10 2/3 | 7 9 | 7 10 15/16 | 8 0 7/8 | 8 2 13/16 |
| 32 | 2133 | 7 1 1/3 | 8 0 | 8 2 | 8 4 | 8 6 |
| 33 | 2200 | 7 4 | 8 3 | 8 5 1/16 | 8 7 1/8 | 8 9 3/16 |
| 34 | 2267 | 8 6 2/3 | 8 6 | 8 8 1/8 | 8 10 1/4 | 9 0 3/8 |
| 35 | 2333 | 7 9 1/3 | 8 9 | 8 11 3/16 | 9 1 3/8 | 9 3 9/16 |
| 36 | 2400 | 8 0 | 9 0 | 9 2 1/4 | 9 4 1/2 | 9 6 3/4 |
| 37 | 2467 | 8 2 2/3 | 9 3 | 9 5 5/16 | 9 7 5/8 | 9 9 15/16 |
| 38 | 2533 | 8 5 1/3 | 9 6 | 9 8 3/8 | 9 10 3/4 | 10 1 1/8 |
| 39 | 2600 | 8 8 | 9 9 | 9 11 7/16 | 10 1 7/8 | 10 4 5/16 |
| 40 | 2667 | 8 10 2/3 | 10 0 | 10 2 1/2 | 10 5 | 10 7 1/2 |
| 41 | 2733 | 9 1 1/3 | 10 3 | 10 5 9/16 | 10 8 1/8 | 10 11 11/16 |
| 42 | 2800 | 9 4 | 10 6 | 10 8 5/8 | 10 11 1/4 | 11 1 7/8 |
| 43 | 2867 | 9 6 2/3 | 10 9 | 10 11 11/16 | 11 2 3/8 | 11 5 1/16 |
| 44 | 2933 | 9 9 1/3 | 11 0 | 11 1 3/4 | 11 5 1/2 | 11 8 1/4 |
| 45 | 3000 | 10 0 | 11 3 | 11 5 13/16 | 11 8 5/8 | 11 11 7/16 |
| 46 | 3067 | 10 2 2/3 | 11 6 | 11 8 7/8 | 11 11 3/4 | 12 2 5/8 |
| 47 | 3133 | 10 5 1/3 | 11 9 | 11 11 15/16 | 12 2 7/8 | 12 5 13/16 |
| 48 | 3200 | 10 8 | 12 0 | 12 3 | 12 6 | 12 9 |
| 49 | 3267 | 10 10 2/3 | 12 3 | 12 6 1/16 | 12 9 1/8 | 13 0 3/16 |

| Course Height | Mm Incr 67mm | Ft and Inches Increments 2 2/3 | Ft and Inches Increments 3 | Ft and Inches Increments 3 1/16 | Ft and Inches Increments 3 1/8 | Ft and Inches Increments 3 3/16 |
|---|---|---|---|---|---|---|
| 50 | 3333 | 11 1 1/3 | 12 6 | 12 9 1/8 | 13 0 1/4 | 13 3 3/8 |
| 51 | 3400 | 11 4 | 12 9 | 13 0 3/16 | 13 3 3/8 | 13 6 9/16 |
| 52 | 3467 | 11 6 2/3 | 13 0 | 13 3 1/4 | 13 6 1/2 | 13 9 3/4 |
| 53 | 3533 | 11 9 1/3 | 13 3 | 13 6 5/16 | 13 9 5/8 | 14 0 15/16 |
| 54 | 3600 | 12 0 | 13 6 | 13 9 3/8 | 14 0 3/4 | 14 4 1/8 |
| 55 | 3667 | 12 2 2/3 | 13 9 | 14 0 7/16 | 14 3 7/8 | 14 7 5/16 |
| 56 | 3733 | 12 5 1/3 | 14 0 | 14 3 1/2 | 14 7 | 14 10 1/2 |
| 57 | 3800 | 12 8 | 14 3 | 14 6 9/16 | 14 10 1/8 | 15 1 11/16 |
| 58 | 3867 | 12 10 2/3 | 14 6 | 14 9 5/8 | 15 1 1/4 | 15 4 7/8 |
| 59 | 3933 | 13 1 1/3 | 14 9 | 15 0 11/16 | 15 4 3/8 | 15 8 1/16 |
| 60 | 4000 | 13 4 | 15 0 | 15 3 3/4 | 15 7 1/2 | 15 11 1/4 |
| 61 | 4067 | 13 6 2/3 | 15 3 | 15 6 13/16 | 15 10 5/8 | 16 2 7/16 |
| 62 | 4133 | 13 9 1/3 | 15 6 | 15 9 7/8 | 16 1 3/4 | 16 5 5/8 |
| 63 | 4200 | 14 0 | 15 9 | 16 0 15/16 | 16 4 7/8 | 16 8 13/16 |
| 64 | 4267 | 14 2 2/3 | 16 0 | 16 4 | 16 8 | 17 0 |
| 65 | 4333 | 14 5 1/3 | 16 3 | 16 7 1/16 | 16 11 1/8 | 17 3 3/16 |
| 66 | 4400 | 14 8 | 16 6 | 16 10 1/8 | 17 2 1/4 | 17 6 3/8 |
| 67 | 4467 | 10 10 2/3 | 16 9 | 17 1 3/16 | 17 5 3/8 | 17 9 9/16 |
| 68 | 4533 | 15 1 1/3 | 17 0 | 17 4 1/4 | 17 8 1/2 | 18 0 3/4 |
| 69 | 4600 | 15 4 | 17 3 | 17 7 5/16 | 17 11 5/8 | 18 3 15/16 |
| 70 | 4667 | 15 6 2/3 | 17 6 | 17 10 3/8 | 18 2 3/4 | 18 7 1/8 |
| 71 | 4733 | 15 9 1/3 | 17 9 | 18 1 7/16 | 18 5 7/8 | 18 10 5/16 |
| 72 | 4800 | 16 0 | 18 0 | 19 4 1/2 | 18 9 | 19 1 1/2 |
| 73 | 4867 | 16 2 2/3 | 18 3 | 18 7 7/16 | 19 0 1/8 | 19 4 11/16 |
| 74 | 4933 | 16 5 1/3 | 18 6 | 18 10 5/8 | 19 3 1/4 | 19 7 7/8 |
| 75 | 5000 | 16 8 | 18 9 | 19 1 11/16 | 19 6 3/8 | 19 11 1/16 |
| 76 | 5067 | 16 10 2/3 | 19 0 | 19 4 3/4 | 19 9 1/2 | 20 2 1/4 |
| 77 | 5133 | 17 1 1/3 | 19 3 | 19 7 13/16 | 20 0 5/8 | 20 5 7/16 |

# Appendix B: Brick Gauge

| Course Height | Mm Incre 80mm | Ft and Inches Increments 3 1/5 | Ft and Inches Increments 3 1/4 | Ft and Inches Increments 3 5/16 | Ft and Inches Increments 3 3/8 | Ft and Inches Increments 3 7/16 |
|---|---|---|---|---|---|---|
| 1 | 80 | 3 1/5 | 3 1/4 | 3 5/16 | 3 3/8 | 3 7/16 |
| 2 | 160 | 6 2/5 | 6 1/2 | 6 5/8 | 6 3/4 | 6 7/8 |
| 3 | 240 | 9 3/5 | 9 3/4 | 9 5/16 | 10 1/8 | 10 15/16 |
| 4 | 320 | 1 0 4/5 | 1 1 | 1 1 1/4 | 1 1 1/2 | 1 1 3/4 |
| 5 | 400 | 1 4 | 1 4 1/4 | 1 4 9/16 | 1 4 7/8 | 1 5 3/16 |
| 6 | 480 | 1 7 1/5 | 1 7 1/2 | 1 7 7/8 | 1 8 1/4 | 1 8 5/8 |
| 7 | 560 | 1 10 2/5 | 1 10 3/4 | 1 11 3/16 | 1 11 5/8 | 2 0 1/16 |
| 8 | 640 | 2 1 3/5 | 2 2 | 2 2 1/2 | 2 3 | 2 3 1/2 |
| 9 | 720 | 2 4 4/5 | 2 5 1/4 | 2 5 13/16 | 2 6 3/8 | 2 6 15/16 |
| 10 | 800 | 2 8 | 2 8 1/2 | 2 11 1/8 | 2 9 3/4 | 2 10 3/8 |
| 11 | 880 | 2 11 1/5 | 2 11 3/4 | 3 0 7/16 | 3 1 1/8 | 3 1 13/16 |
| 12 | 960 | 3 2 2/5 | 3 3 | 3 3 3/4 | 3 4 1/2 | 3 5 1/4 |
| 13 | 1040 | 3 5 3/5 | 3 6 1/4 | 3 7 1/16 | 3 7 7/8 | 3 8 11/16 |
| 14 | 1120 | 3 8 4/5 | 3 9 1/2 | 3 10 3/8 | 3 11 1/4 | 4 0 1/8 |
| 15 | 1200 | 4 0 | 4 0 3/4 | 4 1 11/16 | 4 2 5/8 | 4 3 9/16 |
| 16 | 1290 | 4 3 1/5 | 4 4 | 4 5 | 4 6 | 4 7 |
| 17 | 1360 | 4 6 2/5 | 4 7 1/4 | 4 8 5/16 | 4 9 3/8 | 4 10 7/16 |
| 18 | 1440 | 4 9 3/5 | 4 10 1/2 | 4 11 5/8 | 5 0 3/4 | 5 1 7/8 |
| 19 | 1520 | 5 0 4/5 | 5 1 3/4 | 5 2 15/16 | 5 4 1/8 | 5 5 5/16 |
| 20 | 1600 | 5 4 | 5 5 | 5 6 1/4 | 5 7 1/2 | 5 8 3/4 |
| 21 | 1680 | 5 7 1/5 | 5 8 1/4 | 5 9 9/16 | 5 10 7/8 | 6 0 3/16 |
| 22 | 1760 | 5 10 2/5 | 5 11 1/2 | 6 0 7/8 | 6 2 1/4 | 6 3 5/8 |
| 23 | 1840 | 6 1 3/5 | 6 2 3/4 | 6 4 3/16 | 6 5 5/8 | 6 7 1/16 |
| 24 | 1920 | 6 4 4/5 | 6 6 | 6 7 1/2 | 6 9 | 6 10 1/2 |
| 25 | 2000 | 6 7 | 6 9 1/4 | 6 10 13/16 | 7 0 3/8 | 7 1 15/16 |
| 26 | 2080 | 6 11 1/5 | 7 0 1/2 | 7 2 1/8 | 7 3 3/4 | 7 5 3/8 |
| 27 | 2160 | 7 2 2/5 | 7 3 3/4 | 7 5 7/16 | 7 7 1/8 | 7 8 13/16 |
| 28 | 2240 | 7 5 3/5 | 7 7 | 7 8 3/4 | 7 10 1/2 | 8 0 1/4 |
| 29 | 2320 | 7 8 4/5 | 7 10 1/4 | 8 0 1/16 | 8 1 7/8 | 8 3 11/16 |
| 30 | 2400 | 8 0 | 8 1 1/2 | 8 3 3/8 | 8 5 1/4 | 8 7 1/8 |
| 31 | 2480 | 8 3 1/5 | 8 4 3/4 | 8 6 11/16 | 8 8 5/8 | 8 10 9/16 |
| 32 | 2560 | 8 6 2/5 | 8 8 | 8 10 | 9 0 | 9 2 |
| 33 | 2640 | 8 9 3/5 | 8 11 1/4 | 9 1 5/16 | 9 3 3/8 | 9 5 7/16 |
| 34 | 2720 | 9 0 4/5 | 9 2 1/2 | 9 4 5/8 | 9 6 3/4 | 9 8 7/8 |
| 35 | 2800 | 9 4 | 9 5 3/4 | 9 7 15/16 | 9 10 1/8 | 10 0 5/16 |
| 36 | 2880 | 9 7 1/5 | 9 9 | 9 11 1/4 | 10 1 1/2 | 10 3 3/4 |
| 37 | 2960 | 9 10 2/5 | 10 0 1/4 | 10 2 9/16 | 10 4 7/8 | 10 7 3/16 |
| 38 | 3040 | 10 1 3/5 | 10 3 1/2 | 10 5 7/8 | 10 8 1/4 | 10 10 5/8 |
| 39 | 3120 | 10 4 4/5 | 10 6 3/4 | 10 9 3/16 | 10 11 5/8 | 11 2 1/16 |
| 40 | 3200 | 10 7 | 10 10 | 11 0 1/2 | 11 3 | 11 5 1/2 |
| 41 | 3280 | 10 11 1/5 | 11 1 1/4 | 11 3 13/16 | 11 6 3/8 | 11 8 15/16 |
| 42 | 3360 | 11 2 2/5 | 11 4 1/2 | 11 7 1/8 | 11 9 3/4 | 12 0 3/8 |
| 43 | 3440 | 11 5 3/5 | 11 7 3/4 | 11 10 7/16 | 12 1 1/8 | 12 3 13/16 |
| 44 | 3520 | 11 8 4/5 | 11 11 | 12 1 3/4 | 12 4 1/2 | 12 7 1/4 |
| 45 | 3600 | 12 0 | 12 2 1/4 | 12 5 1/16 | 12 7 7/8 | 12 10 11/16 |
| 46 | 3680 | 12 3 1/5 | 12 5 1/2 | 12 8 3/8 | 12 11 1/4 | 13 2 1/8 |
| 47 | 3760 | 12 6 2/5 | 12 8 3/4 | 12 11 11/16 | 13 2 5/8 | 13 5 9/16 |
| 48 | 3840 | 12 9 3/5 | 13 0 | 13 3 | 13 6 | 13 9 |
| 49 | 3920 | 13 0 4/5 | 13 3 1/4 | 13 6 5/16 | 1 9 3/8 | 14 0 7/16 |

| Course Height | Mm Incre 80mm | Ft and Inches Increments 3 1/5 | Ft and Inches Increments 3 1/4 | Ft and Inches Increments 3 5/16 | Ft and Inches Increments 3 3/8 | Ft and Inches Increments 3 7/16 |
|---|---|---|---|---|---|---|
| 50 | 4000 | 13  4 | 13  6  1/2 | 13  9  5/16 | 14  0  3/4 | 14  3  7/8 |
| 51 | 4080 | 13  7  1/5 | 13  9  3/4 | 14  0  15/16 | 14  4  1/8 | 14  7  5/16 |
| 52 | 4160 | 13  10  2/5 | 14  1 | 14  4  1/4 | 14  7  1/2 | 14  10  3/4 |
| 53 | 4240 | 14  1  3/5 | 14  3  1/4 | 14  7  9/16 | 14  10  7/8 | 15  2  3/16 |
| 54 | 4320 | 14  4  4/5 | 14  6  1/2 | 14  10  7/8 | 15  2  1/4 | 15  5  5/8 |
| 55 | 4400 | 14  8 | 14  9  3/4 | 15  2  3/16 | 15  5  5/8 | 15  9  1/16 |
| 56 | 4480 | 14  11  1/5 | 15  2 | 15  5  1/2 | 15  9 | 16  0  1/2 |
| 57 | 4560 | 15  2  2/5 | 15  5  1/4 | 15  8  13/16 | 16  0  3/8 | 16  3  15/16 |
| 58 | 4640 | 15  5  3/5 | 15  8  1/2 | 16  0  1/8 | 16  3  3/4 | 16  7  3/8 |
| 59 | 4720 | 15  8  4/5 | 15  11  3/4 | 16  3  7/16 | 16  7  1/8 | 16  10  13/16 |
| 60 | 4800 | 16  0 | 16  3 | 16  6  3/4 | 16  10  1/2 | 17  2  1/4 |
| 61 | 4880 | 16  3  1/5 | 16  6  1/4 | 16  10  1/6 | 17  1  7/8 | 17  5  11/16 |
| 62 | 4960 | 16  6  2/5 | 16  9  1/2 | 17  1  3/8 | 17  5  1/4 | 17  9  1/8 |
| 63 | 5040 | 16  9  3/5 | 17  0  3/4 | 17  4  11/16 | 17  8  5/8 | 18  0  9/16 |
| 64 | 5120 | 17  0  4/5 | 17  4 | 17  8 | 18  0 | 18  4 |
| 65 | 5200 | 17  4 | 17  7  1/4 | 17  11  5/16 | 18  3  3/8 | 18  7  7/16 |
| 66 | 5280 | 17  7  1/2 | 17  10  1/2 | 18  2  5/8 | 18  6  3/4 | 18  10  7/8 |
| 67 | 5360 | 17  10  2/5 | 18  1  3/4 | 18  5  15/16 | 18  10  1/8 | 19  2  5/16 |
| 68 | 5440 | 18  1  3/5 | 18  5 | 18  9  1/4 | 19  1  1/2 | 19  5  3/4 |
| 69 | 5520 | 18  4  4/5 | 18  8  1/4 | 19  0  9/16 | 19  4  7/8 | 19  9  3/16 |
| 70 | 5600 | 18  7 | 18  11  1/2 | 19  3  7/8 | 19  8  1/4 | 20  0  5/8 |
| 71 | 5680 | 18  11  1/5 | 19  2  3/4 | 19  7  3/16 | 19  11  5/8 | 20  4  1/16 |
| 72 | 5760 | 19  2  2/5 | 19  6 | 19  10  1/2 | 20  3 | 20  7  1/2 |
| 73 | 5840 | 19  5  3/5 | 19  9  1/4 | 20  1  13/16 | 20  6  3/8 | 20  10  15/16 |
| 74 | 5920 | 19  8  4/5 | 20  0  1/2 | 20  5  1/8 | 20  9  3/4 | 21  2  3/8 |
| 75 | 6000 | 20  0 | 20  3  3/4 | 20  8  7/16 | 21  1  1/8 | 21  5  13/16 |
| 76 | 6080 | 20  3  1/5 | 20  7 | 20  11  3/4 | 21  4  1/2 | 21  9  1/4 |
| 77 | 6160 | 20  5  2/5 | 20  10  1/4 | 21  3  1/16 | 21  7  7/8 | 22  0  11/16 |

# Appendix B: Brick Gauge

| Course Height | Ft and Inches Increments 3½ | Ft and Inches Increments 3 9/16 | Ft and Inches Increments 3 5/8 | Ft and Inches Increments 3 11/16 | Ft and Inches Increments 3 ¾ | Mm Incre 100mm | Ft and Inches Increments 4 |
|---|---|---|---|---|---|---|---|
| 1 | 3 ½ | 3 9/16 | 3 5/8 | 3 11/16 | 3 ¾ | 100 | 4 |
| 2 | 7 | 7 1/8 | 7 ¼ | 7 3/8 | 7 ½ | 200 | 8 |
| 3 | 10 ½ | 10 11/16 | 10 7/8 | 11 1/16 | 11 ¼ | 300 | 1 0 |
| 4 | 1 2 | 1 2 ¼ | 1 2 ½ | 1 2 ¾ | 1 3 | 400 | 1 4 |
| 5 | 1 5 ½ | 1 5 13/16 | 1 6 1/8 | 1 6 7/16 | 1 6 ¾ | 500 | 1 8 |
| 6 | 1 9 | 1 9 3/8 | 1 9 ¾ | 1 10 1/8 | 1 10 ½ | 600 | 2 0 |
| 7 | 2 0 ½ | 2 0 15/16 | 2 1 3/8 | 2 1 13/16 | 2 2 ¼ | 700 | 2 4 |
| 8 | 2 4 | 2 4 ½ | 2 5 | 2 5 ½ | 2 6 | 800 | 2 8 |
| 9 | 2 7 ½ | 2 8 1/16 | 2 8 5/8 | 2 9 3/16 | 2 9 ¾ | 900 | 3 0 |
| 10 | 2 11 | 2 11 5/8 | 3 0 ¼ | 3 0 7/8 | 3 1 ½ | 1000 | 3 4 |
| 11 | 3 2 ½ | 3 3 3/16 | 3 3 7/8 | 3 4 9/16 | 4 5 ¼ | 1100 | 3 8 |
| 12 | 3 6 | 3 6 ¾ | 3 7 ½ | 3 8 ¼ | 3 9 | 1200 | 4 0 |
| 13 | 3 9 ½ | 3 10 15/16 | 3 11 1/8 | 3 11 15/16 | 4 0 ¾ | 1300 | 4 4 |
| 14 | 4 1 | 4 1 7/8 | 4 2 ¾ | 4 3 5/8 | 4 4 ½ | 1400 | 4 8 |
| 15 | 4 4 ½ | 4 5 7/16 | 4 6 3/8 | 4 7 5/16 | 4 8 ¼ | 1500 | 5 0 |
| 16 | 4 8 | 4 9 | 4 10 | 4 11 | 5 0 | 1600 | 5 4 |
| 17 | 4 11 ½ | 5 0 9/16 | 5 1 5/8 | 5 2 11/16 | 5 3 ¾ | 1700 | 5 8 |
| 18 | 5 3 | 5 4 1/8 | 5 5 ¼ | 5 6 3/8 | 5 7 ½ | 1800 | 6 0 |
| 19 | 5 6 ½ | 5 7 11/16 | 5 8 7/8 | 5 10 1/16 | 5 11 ¼ | 1900 | 6 4 |
| 20 | 5 10 | 5 11 ¼ | 6 0 ½ | 6 1 ¾ | 6 3 | 2000 | 6 8 |
| 21 | 6 1 ½ | 6 2 13/16 | 6 4 1/8 | 6 5 7/16 | 6 6 ¾ | 2100 | 7 0 |
| 22 | 6 5 | 6 6 3/8 | 6 7 ¾ | 6 9 1/8 | 6 10 ½ | 2200 | 7 4 |
| 23 | 6 8 ½ | 6 9 15/16 | 6 11 3/8 | 7 0 13/16 | 7 2 ¼ | 2300 | 7 8 |
| 24 | 7 0 | 7 1 ½ | 7 3 | 7 4 ¼ | 7 6 | 2400 | 8 0 |
| 25 | 7 3 ½ | 7 5 1/16 | 7 6 5/8 | 7 8 3/16 | 7 9 ¾ | 2500 | 8 4 |
| 26 | 7 7 | 7 8 5/18 | 7 10 ¼ | 7 11 7/8 | 8 1 ½ | 2600 | 8 8 |
| 27 | 7 10 ½ | 8 0 3/16 | 8 1 7/8 | 8 3 9/16 | 8 5 ¼ | 2700 | 9 0 |
| 28 | 8 2 | 8 3 ¾ | 8 5 ½ | 8 7 ¼ | 8 9 | 2800 | 9 4 |
| 29 | 8 5 ½ | 8 7 5/15 | 8 9 1/8 | 8 10 15/16 | 9 0 ¾ | 2900 | 9 8 |
| 30 | 8 9 | 8 10 7/8 | 9 0 ¾ | 9 2 5/8 | 9 4 ½ | 3000 | 10 0 |
| 31 | 9 0 ½ | 9 2 7/16 | 9 4 3/8 | 9 6 5/16 | 9 8 ¼ | 3100 | 10 4 |
| 32 | 9 4 | 9 6 | 9 8 | 9 10 | 10 0 | 3200 | 10 8 |
| 33 | 9 7 ½ | 9 9 9/16 | 9 11 5/8 | 10 1 11/16 | 10 3 ¾ | 3300 | 11 0 |
| 34 | 9 11 | 10 1 1/8 | 10 3 ¼ | 10 5 3/8 | 10 7 ½ | 3400 | 11 4 |
| 35 | 10 2 ½ | 10 4 11/16 | 10 6 7/8 | 10 9 1/16 | 10 11 ¼ | 3500 | 11 8 |
| 36 | 10 6 | 10 8 ¼ | 10 10 ½ | 11 0 ¾ | 11 3 | 3600 | 12 0 |
| 37 | 10 9 ½ | 10 11 13/16 | 11 2 1/8 | 11 4 7/16 | 11 6 ¾ | 3700 | 12 4 |
| 38 | 11 1 | 11 3 3/8 | 11 5 ¾ | 11 8 1/8 | 11 10 ½ | 3800 | 12 8 |
| 39 | 11 4 ½ | 11 6 15/16 | 11 9 3/8 | 11 11 13/16 | 12 2 ¼ | 3900 | 13 0 |
| 40 | 11 8 | 11 10 ½ | 12 1 | 12 3 ½ | 12 6 | 4000 | 13 4 |
| 41 | 11 11 ½ | 12 2 1/16 | 12 4 5/8 | 12 7 3/16 | 12 9 ¾ | 4100 | 13 8 |
| 42 | 12 3 | 12 5 5/8 | 12 8 ¼ | 12 10 7/8 | 13 1 ½ | 4200 | 14 0 |
| 43 | 12 6 ½ | 12 9 3/16 | 12 11 7/8 | 13 2 9/16 | 13 5 ¼ | 4300 | 14 4 |
| 44 | 12 10 | 13 0 ¾ | 3 3 ½ | 13 6 ¼ | 13 9 | 4400 | 14 8 |
| 45 | 13 1 ½ | 13 4 5/16 | 13 7 1/8 | 13 9 15/16 | 14 0 ¾ | 4500 | 15 0 |
| 46 | 13 5 | 13 7 7/8 | 13 10 ¾ | 14 1 5/8 | 14 4 ½ | 4600 | 15 4 |
| 47 | 13 8 ½ | 13 11 7/16 | 14 2 3/8 | 14 5 5/16 | 14 8 ¼ | 4700 | 15 8 |
| 48 | 14 0 | 14 3 | 14 6 | 14 9 | 15 0 | 4800 | 16 0 |

| Course Height | Ft and Inches Increments 3½ | | | Ft and Inches Increments 3 9/16 | | | Ft and Inches Increments 3 5/8 | | | Ft and Inches Increments 3 11/16 | | | Ft and Inches Increments 3 ¾ | | | Mm Incre 100mm | Ft and Inches Increments 4 | |
|---|---|---|---|---|---|---|---|---|---|---|---|---|---|---|---|---|---|---|
| 49 | 14 | 3 | ½ | 14 | 6 | 9/16 | 14 | 9 | 5/8 | 15 | 0 | 11/16 | 15 | 3 | ¾ | 4900 | 16 | 4 |
| 50 | 14 | 7 | | 14 | 10 | 1/8 | 15 | 1 | ¼ | 15 | 4 | 3/8 | 15 | 7 | ½ | 5000 | 16 | 8 |
| 51 | 14 | 10 | ½ | 15 | 1 | 11/16 | 15 | 4 | 7/8 | 15 | 8 | 1/16 | 15 | 11 | ¼ | 5100 | 17 | 0 |
| 52 | 15 | 2 | | 15 | 5 | ¼ | 15 | 8 | ½ | 15 | 11 | ¾ | 16 | 3 | | 5200 | 17 | 4 |
| 53 | 15 | 5 | ½ | 15 | 8 | 13/16 | 16 | 0 | 1/8 | 16 | 3 | 7/16 | 16 | 6 | ¾ | 5300 | 17 | 8 |
| 54 | 15 | 9 | | 16 | 0 | 3/8 | 16 | 3 | ¾ | 16 | 7 | 1/8 | 16 | 10 | ½ | 5400 | 18 | 0 |
| 55 | 16 | 0 | ½ | 16 | 3 | 15/16 | 16 | 7 | 3/8 | 16 | 10 | 13/16 | 17 | 2 | ¼ | 5500 | 18 | 4 |
| 56 | 16 | 4 | | 16 | 7 | ½ | 16 | 11 | | 17 | 2 | ½ | 17 | 6 | | 5600 | 18 | 8 |
| 57 | 16 | 7 | ½ | 16 | 11 | 1/16 | 17 | 2 | 5/8 | 17 | 6 | 3/16 | 17 | 9 | ¾ | 5700 | 19 | 0 |
| 58 | 16 | 11 | | 17 | 2 | 5/8 | 17 | 6 | ¼ | 17 | 9 | 7/8 | 18 | 1 | ½ | 5800 | 19 | 4 |
| 59 | 17 | 2 | ½ | 17 | 6 | 3/16 | 17 | 9 | 7/8 | 18 | 1 | 9/16 | 18 | 5 | ¼ | 5900 | 19 | 8 |
| 60 | 17 | 6 | | 19 | 9 | ¾ | 18 | 1 | ½ | 18 | 5 | ¼ | 18 | 9 | | 6000 | 20 | 0 |
| 61 | 17 | 9 | ½ | 18 | 1 | 5/16 | 18 | 5 | 1/8 | 18 | 8 | 15/16 | 19 | 0 | ¾ | 6100 | 20 | 4 |
| 62 | 18 | 1 | | 18 | 4 | 7/8 | 18 | 8 | ¾ | 19 | 0 | 5/8 | 19 | 4 | ½ | 6200 | 20 | 8 |
| 63 | 18 | 4 | ½ | 18 | 8 | 7/16 | 19 | 0 | 3/8 | 19 | 4 | 5/16 | 19 | 8 | ¼ | 6300 | 21 | 0 |
| 64 | 18 | 8 | | 19 | 0 | | 19 | 4 | | 19 | 8 | | 20 | 0 | | 6400 | 21 | 4 |
| 65 | 18 | 11 | ½ | 19 | 3 | 9/16 | 19 | 7 | 5/8 | 19 | 11 | 11/16 | 20 | 3 | ¾ | 6500 | 21 | 8 |
| 66 | 19 | 3 | | 19 | 7 | 1/8 | 19 | 11 | ¼ | 20 | 3 | 3/8 | 20 | 7 | ½ | 6600 | 22 | 0 |
| 67 | 19 | 6 | ½ | 19 | 10 | 11/16 | 20 | 2 | 7/8 | 20 | 7 | 1/16 | 20 | 11 | ¼ | 6700 | 22 | 4 |
| 68 | 19 | 10 | | 20 | 2 | ¼ | 20 | 6 | ½ | 20 | 10 | ¾ | 21 | 3 | | 6800 | 22 | 8 |
| 69 | 20 | 1 | ½ | 20 | 5 | 13/16 | 20 | 10 | 1/8 | 21 | 2 | 7/16 | 21 | 6 | ¾ | 6900 | 23 | 0 |
| 70 | 20 | 5 | | 20 | 9 | 3/8 | 21 | 1 | ¾ | 21 | 6 | 1/8 | 21 | 10 | ½ | 7000 | 23 | 4 |
| 71 | 20 | 8 | ½ | 21 | 0 | 15/16 | 21 | 5 | 3/8 | 21 | 9 | 13/16 | 22 | 2 | ¼ | 7100 | 23 | 8 |
| 72 | 21 | 0 | | 21 | 4 | ½ | 21 | 9 | | 22 | 1 | ½ | 22 | 6 | | 7200 | 24 | 0 |
| 73 | 21 | 3 | ½ | 21 | 8 | 1/16 | 22 | 0 | 5/8 | 22 | 5 | 3/16 | 22 | 9 | ¾ | 7300 | 24 | 4 |
| 74 | 21 | 7 | | 21 | 11 | 5/8 | 22 | 4 | ¼ | 22 | 8 | 7/8 | 23 | 1 | ½ | 7400 | 24 | 8 |
| 75 | 21 | 10 | ½ | 22 | 3 | 3/16 | 22 | 7 | 7/8 | 23 | 0 | 9/16 | 23 | 5 | ¼ | 7500 | 25 | 0 |
| 76 | 22 | 2 | | 22 | 6 | ¾ | 22 | 11 | ½ | 23 | 4 | ¼ | 23 | 9 | | 7600 | 25 | 4 |
| 77 | 22 | 5 | ½ | 22 | 10 | 5/16 | 23 | 3 | 1/8 | 23 | 7 | 15/16 | 24 | 0 | ¾ | 7700 | 25 | 8 |

# Glossary

**Ant capping**  Designed as a slow-up to ants (termites) which travel up from their ground dwelling to feed on timber

**Arch**  The special arrangement of bricks to span an opening

**Arch bar**  Same as a *Lintel bar*

**Arris**  The edge or angle of the brick

**Bat**  Part of a brick greater than one quarter

**Batter board**  A shaped or tapered board used in setting out the batter of a wall (Also referred to as a *template*.)

**Bay windows**  Distinctive style feature used in many modern houses whereby window has three faces and juts out from the contour of the house

**Bed**  (a) The underside of a brick

(b) Spread mortar upon which the brick or piece of construction is laid or bedded

**Bedding**  The process of laying in position a brick, stone, or piece of construction

**Bed joint**  The mortar joint which is horizontal on the face of the wall

**Block sash**  A block that is specially grooved to carry the fin of a window and a door

**Bond**  The arrangement of bricks overlapping one another in a definite pattern, at the same time maintaining the greatest possible strength

**Brick veneer**  Where a timber frame is enclosed by a skin of bricks, and a cavity separates the two walls

**Broken bond**  Occurs where the number of bricks will not fit exactly into the required length of a wall, so that a brick of odd size has to be inserted in each course

**Bull nose**  A rounded front edge

**Butt**  To place against, without overlapping

**Buttering**  Process of placing mortar on the end or face of a brick, whilst held in the hand

**Buttress**  An attached pier designed to counteract a side thrust

| | |
|---|---|
| **Cap** | The top cover of a pier or chimney |
| **Cavity bricks** | Two skins of brickwork with a space between |
| **Chase** | A narrow recess cut in the brickwork (to take a conduit or piping) |
| **Chimney rag** | A piece of viscreen with ropes attached that can be removed on completion of the stack, completely removing any droppings from the smoke shelf |
| **Chimney breast** | The projecting portion of a chimney which contains the flues |
| **Chimney stack** | The portion of the chimney construction containing the upper sections of the flues which passes through and projects above the roof |
| **Chimney throat** | The entrance to the flue from the fireplace |
| **Closures** | Bricks cut to sizes, generally 1¾" (45 mm) for the bonding and tying of walls and piers |
| **Compo** | A mixture of sand, cement, and lime mortar |
| **Coping** | The covering feature of an external wall |
| **Corbel** | A support projecting from the face of a wall |
| **Corbelling** | Building out from the face of a wall in successive projecting courses |
| **Course** | A row of bricks which extends the full length of the wall |
| **Crown** | The top of an arch |
| **Damper** | Device, usually made from stainless steel, designed to prevent the heat in a fireplace from escaping up the chimney |
| **Damp proof course (DPC)** | A waterproof membrane inserted in brickwork to prevent the passage of moisture |
| **Datum** | A level point to work from |
| **Dead man** | A temporary intermediate pier usually constructed toward the middle of a wall to support a string line |
| **English bond** | This bond is sometimes referred to as old English bond. It is recognised as being the strongest of all bonds used in brickwork. |
| **Expansion joint** | A vertical or horizontal joint placed in brick construction to allow for expansion or brick growth |
| **Extrados** | The upper surface of an arch |
| **Extruded brick** | Bricks with a series of holes |

| | |
|---|---|
| **Fender wall** | A dwarf wall to carry the hearth of a ground floor fireplace |
| **Fireplace jambs** | The brick piers forming the sides of the fireplace opening |
| **Flemish bond** | This bond consists of laying one stretcher and one header alternately in the same course. It is not as strong as English bond but is considered by some authorities to present a more pleasing appearance than English bond. |
| **Flue** | A pipe or tube formed for conveying smoke or air |
| **Frog brick** | The indentation in the bedding face of the brick. It helps to reduce the weight of the brick and provides a key for the mortar joint. The real purpose of a frog is primarily to force the brick material into the corners of the mold when manufacturing dry-press bricks. |
| **Gathering** | (a) The reduction of the brickwork opening over the mouth of the fireplace so as to reduce to the required size of the flue |
| | (b) Bringing together all the flues to the base of the stack |
| | (c) Any part of the flue that changes direction |
| **Gauge** | Combined measurements of brick and mortar |
| **Half bat** | Commonly referred to as a bat and is half the length of the brick |
| **Haunch** | Shoulder of an arch; the portion between the springing line and the crown |
| **Head** | Horizontal top members of an opening, or of a door or window frame |
| **Header** | Bricks laid to show a 3⅝" × 2¼" (90 × 57 mm) on the face of the wall |
| **Header bond** | This bond is not used to any great extent in normal brick construction, but it is particularly useful for walls which have to be curved. The bond would always be 1¾" (45 mm). When a sharp curve is required the bricks will require cutting. When the curve is not sharp and the wall has to be 7⅝" (190 mm) thick, a gradual curve can be constructed, resulting in wedge-shaped joints. Curved walls using header bond can also be constructed when the wall thickness is 3⅝" (90 mm). In this case snap headers or bats are used. |
| **Hearth** | The slab projecting in front of the fireplace opening and jambs |
| **Indents** | This method is used when a wall has to be engaged into an existing wall, when the wall is 3⅝" (90 mm) single skin. King |

closures should be used to help retain the strength and looks of the wall and to prevent the cracking of the bat or straight joint. In some cases there might have to be more than one course left between each indent.

**Intrados**          The underside or soffit of an arch

**Jamb**              The brickwork on either side of an opening

**King closure**      A triangular portion of the brick is beveled so as to obtain a 1¾" (45 mm) width at one end and a 3⅝" (90 mm) width at the other end.

**Lintel**            A horizontal member for spanning an opening

**Lipping**           Where the overlapping brickwork protrudes in or out, having a lipping effect

**Modules**           A specific length of brick

**Mortar**            Mixture of sand, cement, lime, and plasticizer, to set requirements

**Mouth of flue**     The entrance to the flue above the fireplace opening

**Nib**               A small pointed projection

**Parging**           Rendering the inside of a flue to give a smooth surface

**Perpends**          Perpends or cross joints in face brickwork should be perpendicular. This is not only because the utmost strength is obtained, but also, and this is equally important, because of the appearance of the work. Perpends in face brickwork which are staggered unevenly up the face of the wall present a most unsightly appearance.

**Pier**              A pillar of brickwork to form a support

**Pointing**          The finishing of the joints of brickwork by pressing in the jointing mortar with the trowel or other tool

**Pressed brick**     Another name for *Frog brick*

**Profiles**          Boards fixed horizontally on pegs driven into the ground (also known as hurdles); alternative method for the building of corners

**Queen closure**     It is commonly referred to as closure. A portion of a brick 1¾" (45 mm) wide. Closures may be the full length of a brick 7⅝" (190 mm), or half the length of a brick 3⅝" (90 mm). Closures may be obtained direct from the manufacturer or cut when required from bricks on the job.

| | |
|---|---|
| **Quoin** | The vertical external angle of a wall |
| **Quoin bricks** | The face bricks forming the external angle of the wall |
| **Racking** | This is the term applied to the process of building up a corner prior to the completion of the remaining portion of the wall |
| **Racking back** | Setting back successive courses of bricks for building up to a later stage |
| **Radius rod** | A lath or batten used to mark out a circular curve by being pivoted at one end |
| **Rebate** | A set-back in an opening |
| **Recess** | Total floor area of a fireplace |
| **Retaining wall** | A wall built for the purpose of keeping in position a volume of earth or liquid |
| **Returned end** | That section of a wall which changes direction |
| **Reveal** | The projecting portion of a wall at a window or door opening when the jambs are recessed or rebated |
| **Rise of an arch** | The vertical height between the springing line and the crown of an arch |
| **Riser** | The amount of height that steps move up in one step, usually two bricks or 5½" (133 mm) |
| **Rough gauged arch** | Common arch constructed for structural purposes only—not a face brick arch. |
| **Screeding course** | Course of common bricks stepped in the distance of the cavity, and for easy removal butted together with no perp joints; removed when the floor is poured |
| **Skew backs** | The bricks forming the sloping surface of an abutment at the springing point of an arch |
| **Soffit** | Same as *Intrados* |
| **Soldier course** | Courses of bricks laid vertically on end |
| **Span** | The distance across an opening |
| **Spiral board** | Piece of board cut to size of pier, with a hole drilled through the center (to fit over cyclone rod) |

| | |
|---|---|
| **Springing points** | Are at the bottom of the intrados of an arch at each end of the springing line |
| **Squint quoin** | A salient angle or corner of a wall, other than a right angle |
| **Stack** | See *Chimney stack* |
| **Stopped end** | The extreme end of a wall which is stopped and does not change direction |
| **Stretcher** | Bricks laid to show a 7⅝" × 2¼" (190 × 57 mm) face on the wall |
| **Stretcher bond** | This bond is only used in walls which are 3⅝" (90 mm) thick, and consists of all stretchers laid with a 3⅝" (90 mm) lap. This is sometimes referred to as half bond, because the overlap is half the length of the brick. |
| **Stanchion** | An upright reinforcement that gives support to walls |
| **Template (or arch center)** | Form of material used to temporarily support an arch, or in scribing, as a batter board |
| **Tingle** | Is cut from a piece of flat metal and is used to hold the line plumb and correct the sag in the line |
| **Toothing** | This is used when some future extension is intended. It should never be used for any other reason unless absolutely unavoidable. Toothing is carried out by leaving each alternate brick projecting beyond the course below. This allows new work to be built into the old. |
| **Trammel** | Essential tool that can save building templates for arches, round windows, bullseyes, etc.; can be made simply from pipe, G-clamps, and a pivot joint |
| **Voussoirs** | Wedge-shaped bricks or blocks of which a built-up arch is composed |
| **Weep holes** | Small openings in a wall to permit the escape of water from the back of the wall |
| **Window fin** | Form of metal protruding from the frame to aid in holding the frame rigid; also helps waterproofing, and is a means to fix and tie the window frame |
| **Withes** | Narrow partitions between flues or pockets |

# Index

# ABOUT THE AUTHOR

**Peter Cartwright** of Queensland, Australia, completed a two-year apprenticeship program in 1975 that required residential, high-rise, ornamental, arch and church, and commercial construction mastery. He earned the title Apprentice of the Year in 1975 and upon completion of the program began his own business which continues to this day.